A Modular Framework for Optimizing Grid Integration of Mobile and Stationary Energy Storage in Smart Grids

Dominik Pelzer

A Modular Framework for Optimizing Grid Integration of Mobile and Stationary Energy Storage in Smart Grids

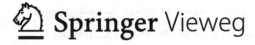

Springer Vieweg

Dominik Pelzer
Hamburg, Germany

Dissertation TU München, Germany, 2018

ISBN 978-3-658-27023-0   ISBN 978-3-658-27024-7   (eBook)
https://doi.org/10.1007/978-3-658-27024-7

Springer Vieweg

This Springer Vieweg imprint is published by the registered company Springer Fachmedien Wiesbaden GmbH part of Springer Nature
The registered company address is: Abraham-Lincoln-Str. 46, 65189 Wiesbaden, Germany

Table of Contents

List of Figures

List of Tables

List of Listings

Nomenclature

B2G	Battery-to-Grid
BESS	Battery Energy Storage System
BMS	Battery Management System
BOL	Beginning of Life
C-rate	Charge Rate
CC	Constant Current
CityMoS	City Mobility Simulation
CPP	Critical Peak Pricing
CV	Constant Voltage
DES	Discrete Event Simulation
DES	Distributed Energy Storage
DOD	Depth of Discharge
DP	Dynamic Programming
DSM	Demand-Side Management
EC	Equivalent Circuit
EOL	End of Life
ESS	Energy Storage System
ET	Electricity Tariff
FERC	Federal Energy Regulatory Commission
G2B	Grid-to-Battery
G2S	Grid-to-Storage
HESS	Hybrid Energy Storage System
HITS	Household Interview Travel Survey
HPC	High Performance Computing
HV	High Voltage
ICEV	Internal Combustion Engine Vehicle
IESO	Independent Electricity System Operator
ISO	Independent System Operator
LMP	Locational Marginal Price
LV	Low Voltage
MCP	Market Clearing Price
MINLP	Mixed Integer Non-Linear Program
MPC	Model Predictive Control
MV	Medium Voltage

NEMS	National Electricity Market of Singapore
NMP	Nodal Marginal Pricing
NYISO	New York Independent System Operator
OCV	Open Circuit Voltage
PEV	Plug-In Electric Vehicle
PHEV	Plug-In Hybrid Electric Vehicle
PJM	Pennsylvania-New Jersey-Maryland Interconnection
PNM	Power Network Model
PSO	Power System Operator
PV	Photovoltaic
RES	Renewable Energy Sources
ROI	Return on Investment
RTO	Regional Transmission Organization
S2G	Storage-to-Grid
SEI	Solid Electrolyte Interface
SEMSim	Scaleable Electric Mobility Simulation
SOC	State of Charge
SOH	State of Health
TCO	Total Cost of Ownership
TOU	Time-of-Use
TSO	Transmission System Operator
UCTE	Union for the Coordination of the Transmission of Electricity
USEP	Uniform Singapore Electricity Price
V2G	Vehicle-to-Grid
VPP	Virtual Power Plant
XSD	XML Schema Definition
ZMP	Zonal Marginal Pricing

Abstract

Distributed *energy storage systems* (ESSs) including *plug-in electric vehicle* (PEV) batteries and stationary small-scale *battery energy storage systems* (BESSs) have the potential to play a major role in the transition towards a smart grid. On the one hand, the grid integration of PEVs poses challenges to power system operators as uncoordinated charging may lead to demand peaks and grid overloading. On the other hand, however, PEVs as well as distributed stationary BESSs can be instrumental in reducing the emission intensity of electricity generation through facilitating the grid integration of *renewable energy sources* (RES). By making intelligent use of their storage capacity, ESSs can turn the power system from a just-in-time inventory system in which energy has to be generated precisely at the time it is needed into a system where power demand and supply are decoupled from each other. This allows increasing the share of intermittent RES without compromising power system stability.

While the concept of deploying distributed ESSs for balancing power demand and supply has received growing attention in recent years, its economic viability especially with regard to battery storage is still subject to controversial discussion. In most cases, it is argued that battery wear overcompensates the revenues BESSs could attain by providing services to the power grid, thus rendering the concept financially unviable. Due to the large number of involved parameters, the variety of battery technologies and the dynamics of battery prices over time, generalized conclusions on financial viability are, however, hard to make. This work therefore presents a modular and extensible framework with the primary purpose of systematically investigating the financial aspects of grid-integrated ESSs. Due to its modular architecture, the framework can be modified and extended by adapting, extending or adding individual models to account for the large variety of different types of ESSs, system components, parameter values and market conditions. This allows tailoring the simulation environment to the actual problem in order to identify optimal system design parameters and to compare different scenarios with each other. The scheduling approach which is part of the framework implements a *model predictive control* (MPC) concept which optimizes the charging and dispatching of energy by considering real-time electricity prices, price forecasts, availability constraints and storage system behavior. In the presented context, a focus is placed on the aspect of battery wear and its appropriate consideration and monetization as part of optimal charge/dispatch scheduling since this aspect is identified to be of major importance for profitable BESS operation.

The developed framework is deployed to assess the financial benefits of PEV smart charging for the case of Singapore and to investigate the financial viability of energy arbitrage using BESSs in various electricity markets. The investigation on smart charging performed in conjunction with the traffic simulation CityMoS Traffic shows that the consideration of battery degradation results in considerably higher savings than simply accounting for electricity price fluctuations. The results indicate that under beneficial conditions, PEVs can be economically competitive with internal combustion engine vehicles for a significant share of the population. The investigation of the profitability of energy arbitrage in different electricity markets partly confirms concerns about the profitability of BESSs for power grid services. This is shown to be especially true if charge scheduling does not explicitly account for battery wear. The results, however, also show that by appropriately considering both calendar and cycle battery aging as part of the scheduling approach, profitable operation is in fact possible. Particularly for the growing market of residential batteries for buffering generation from photovoltaics, energy arbitrage may contribute to faster amortization of the investment. Profitability can be tweaked by appropriately considering the particular market conditions including the temporal resolution of prices and by choosing parts of the grid with prevailing higher prices and large price fluctuations. This can be a reasonable starting point for niche applications of small-scale BESSs until battery prices have fallen to a sufficiently low level for location-independent mass deployment. For this purpose, the developed framework provides a tool which can be used for conducting substantiated profitability analyses ahead of making any investment decision.

Zusammenfassung

Elektrofahrzeuge und Batteriespeicher können im Hinblick auf die Entwicklung eines Smart Grids eine entscheidende Rolle einnehmen. Auf der einen Seite stellt die Netzintegration von Elektrofahrzeugen eine Herausforderung für Netzbetreiber dar, da unkoordiniertes Laden zu Nachfragespitzen führen und insbesondere in den Verteilnetzen in Versorgungsengpässen resultieren kann. Auf der anderen Seite können Elektrofahrzeuge sowie verteilte stationäre Energiespeicher zur Reduzierung der Emissionen in der Stromerzeugung beitragen, indem sie die Netzintegration von erneuerbaren Energieträgern erleichtern. Dies kann durch die intelligente Nutzung der Kapazität von verteilten Energiespeichern bewerkstelligt werden, die eine Entkopplung von Stromangebot und -nachfrage erlaubt. Damit können größere Anteile erneuerbarer Energieträger mit fluktuierender Erzeugung in die Netze integriert werden, ohne dabei die Netzstabilität zu gefährden.

Während der Ansatz, verteilte stationäre Energiespeicher und die Batterien von Elektrofahrzeugen zur Netzstabilisierung einzusetzen in den letzten Jahren zunehmend an Aufmerksamkeit gewonnen hat, wird die wirtschaftliche Machbarkeit des Konzepts insbesondere im Hinblick auf Batteriespeicher nach wie vor kontrovers diskutiert. In vielen Fällen wird dabei argumentiert, dass Einnahmen, die durch die Bereitstellung von Services für die Netzbetreiber generiert werden können, durch Batterieabschreibungskosten überkompensiert werden. Allgemeingültige Aussagen zu dieser Thematik sind aufgrund der großen Anzahl von Parametern, Batterietypen und dem fortschreitenden Preisverfall von Batterien jedoch schwer möglich. Deshalb wird in der vorliegenden Arbeit ein modulares Simulationsframework vorgestellt, welches die systematische Untersuchung finanzieller Aspekte verschiedener Typen von Energiespeichern ermöglicht. Aufgrund seiner modularen Architektur kann das Framework angepasst oder erweitert werden, um somit der Vielfalt von Speichertechnologien, Systemkomponenten, Parametern und Marktbedingungen Rechnung zu tragen. Damit kann die Simulation auf die zu untersuchende Problemstellung zugeschnitten werden, um optimale Systemdesignparameter zu identifizieren und verschiedene Szenarien miteinander zu vergleichen. Die *Schedulingmethode* implementiert als Teil des Frameworks einen *Model Predictive Control* Ansatz, der das Laden und Entladen eines Energiespeichers unter Berücksichtigung von Echtzeitinformation zu Strompreisen, Preisvorhersagen, Verfügbarkeitsbeschränkungen und Verhalten des Speichersystems optimiert. In diesem Zusammenhang liegt ein Schwerpunkt auf der expliziten Berücksichtigung und Monetarisierung von Batteriealterungsprozessen, da diesen für den profitablen Betrieb von Batteriespeichern eine entscheidende Bedeutung zukommt.

Das entwickelte Framework wird im Rahmen dieser Arbeit zur Untersuchung des finanziellen Nutzens von intelligentem Laden von Elektrofahrzeugen für den Fall von Singapur und für die Untersuchung der Wirtschaftlichkeit von Energiearbitrage mit Batteriespeichern in verschiedenen Märkten eingesetzt. Die Untersuchung von intelligenten Ladestrategien, die unter zusätzlicher Verwendung der Verkehrssimulation CityMoS Traffic durchgeführt wird, zeigt, dass die Berücksichtigung von Batteriealterung für optimales Laden die Vorteile von preissensitivem Laden deutlich übersteigen kann. Aus den Ergebnissen kann geschlossen werden, dass unter günstigen Umständen und unter Anwendung von intelligenten Ladestrategien Elektrofahrzeuge für einen nennenswerten Teil der Bevölkerung Kostenvorteile im Vergleich zu Verbrennungsfahrzeugen bieten können. Die Untersuchung der Profitabilität von Energiearbitrage in verschiedenen Strommärkten bestätigt in Teilen die Bedenken, dass der Einsatz von Batterieenergiespeichern für Services für Stromnetzbetreiber nicht profitabel ist. Dies trifft insbesondere dann zu, wenn Batterieabnutzung bei der Lade- und Entladesteuerung nicht berücksichtigt wird. Die Ergebnisse zeigen jedoch auch, dass die explizite Berücksichtigung von Batteriealterung als Teil des Schedulings einen profitablen Betrieb ermöglichen kann. Insbesondere für den wachsenden Markt von Heimenergiespeichern zur Kombination mit Photovoltaikanlagen könnte Energiearbitrage zur schnelleren Amortisierung der Investition beitragen. Die Profitabilität kann durch die Berücksichtigung der Strommarktbedingungen insbesondere durch die Wahl von Zonen im Stromnetz mit hoher Preisvariabilität optimiert werden. Dies kann einen Startpunkt für Nischenapplikationen darstellen, bis die Batteriepreise das Niveau für einen ortsunabhängigen Masseneinsatz erreicht haben.

1 Introduction

1.1 Motivation

Commitments for reducing greenhouse gas emissions, the aim for greater independence from fossil fuels and the need to improve urban air quality require fundamental changes to power systems and transportation infrastructures. The main measures in this context include a transition from *centralized* power generation based on fossil fuels towards a *decentralized* system built on *renewable energy sources* (RES). This should further go along with a decarbonization of road transport based on electric propulsion technologies such as *plug-in electric vehicles* (PEVs). This transition towards RES and the introduction of PEVs pose both challenges and opportunities to various kinds of stakeholders. The main challenges in this regard can be summarized as follows:

- Due to the intermittent character of RES, *power system operators* (PSOs) have to establish new solutions for ensuring power system stability. In order to balance fluctuations of energy production by RES, generation overcapacities are required which are scheduled in accordance with the current demand-supply situation. Due to dispatching based upon merit orders, these generation capacities generally have low utilization rates which result in high capital costs. The continuous ramping up and down causes wear and tear to the equipment and generators are often forced to operate outside their range of maximum efficiency. With increasing shares of RES, the need for balancing fluctuations is likely to further increase. Therefore, the cost of maintaining power system stability can be expected to increase should the use of conventional technical solutions for balancing demand and supply persist.

- Even though the electrification of road transport is largely desirable from an environmental point of view, it also raises new chal-

© Springer Fachmedien Wiesbaden GmbH, part of Springer Nature 2019
D. Pelzer, *A Modular Framework for Optimizing Grid Integration of Mobile and Stationary Energy Storage in Smart Grids*, https://doi.org/10.1007/978-3-658-27024-7_1

lenges to PSOs. With PEVs, a direct link between transportation
and power infrastructure is established due to the fact that PEVs
are part of the transportation system while drawing energy from
the power system. Due to their mobile nature and their high power
rating, PEVs introduce a new type of dynamic load into the power
system. This may result in load peaks posing another threat for
power system stability, in particular with regard to the distribution system.

Despite these challenges, there are also potential synergies between
transport electrification and RES. The first reason for this is that
PEV batteries can serve as distributed *energy storage systems* (ESSs)
for buffering excess energy from RES while feeding energy back into
the grid at times of supply shortages. The second reason is that efforts at making PEVs marketable have led to a considerable decrease
of battery manufacturing costs which in turn makes the installation of
small-scale residential *battery energy storage systems* (BESSs) buffering local RES generation increasingly affordable. The intelligent deployment of these mobile and stationary BESSs can therefore play a
major role towards the development of a smart grid.

As listed below, there are various stakeholders who stand to benefit
from the intelligent deployment of BESSs:

- **PSOs and power utilities**
  The fundamental function of any type of ESS is to decouple power
  generation and electrical load. Batteries have the capability of responding virtually instantaneously to control signals which makes
  them suitable for the most costly services for balancing power system fluctuations. As small-scale BESSs can be distributed over
  the whole network, they allow the storage of energy at locations
  where it is produced or consumed, thus saving transmission costs.
  Distributed BESSs can therefore reduce the need for investments
  in grid expansion, mitigate grid congestion and decrease wear on
  generation facilities for the benefit of PSOs and power utilities.

- **BESS owners**
  Providing services to the power system can generate revenues for owners of mobile and stationary BESSs, thus accelerating the amortization of their investment. This can decrease the *total cost of ownership* (TCO) of PEVs and of systems combining *photovoltaic* (PV) modules with small-scale batteries. This can lead to a wider adoption of those technologies which may in turn result in further decreasing consumer prices as a result of economies of scale and experience curve effects.

- **Society**
  The greater capability of power grids to accommodate RES may lead to a faster reduction of greenhouse gas emissions. At the same time, the reduced need for grid expansion and generation overcapacities could ultimately translate into lower electricity prices.

Despite the possible advantages, there are a number of challenges inhibiting the wide adoption of distributed BESSs:

- **Financial aspects**
  The financial viability of deploying small-scale BESSs for power grid services is still subject to controversial discussion. General conclusions on profitability of BESSs are hard to make because of the large number of parameters and the variety of different system properties involved, some of which can be influenced by the BESS controller while others are externally fixed. On the cost side, relevant parameters include battery prices, charging/discharging efficiencies and battery lifetimes. Especially the latter aspect which is related to battery degradation during operation can play a major role with regard to operating costs. Not appropriately accounting for this aspect is therefore likely to result in accelerated depreciation which has repeatedly been demonstrated to outweigh achievable revenues [1–3]. Since accelerated battery wear is already considered one of the concerns inhibiting the purchase of PEVs [4], this could also be a behavioral barrier for preventing system owners to make their batteries available for providing

power grid services. An aspect concerning the revenue side refers
to electricity prices. As the revenue generated by a BESS stems
from the spread between the prices for buying and selling energy
or from the remuneration for providing capacity, sufficiently high
price levels and price variability are required to compensate operating costs. Prices can notably vary among different markets
and even within different load zones of a single market so that
generalized conclusions on profitability are hard to make.

- **Data protection**
  In order to optimally operate a BESS, different types of information are required about the system as well as about the system
  owner. In the case of a PEV, the battery's primary purpose is
  to provide energy for driving. The provision of power grid services should not interfere with this primary purpose so that the
  charge scheduling mechanism needs to ensure that the battery is
  sufficiently charged when needed. This requires highly detailed
  information on a driver's mobility behavior. In the case of a residential BESS which has the primary function of buffering local
  PV generation and powering residential loads, the battery's *state
  of charge* (SOC) needs to be managed in a way that allows the
  fulfillment of these purposes. This requires data on households'
  load patterns which may in turn reveal certain information on
  the residents. In both cases, system owners may be unwilling to
  share this information for privacy reasons. Furthermore, in order
  to ensure optimal scheduling, e.g., with regard to battery degradation, technical information on the BESS is also required. This
  includes models on the battery behavior at a cell level which manufacturers are likely to be hesitant to share in order to protect their
  intellectual property. Operating a BESS without the specified information may lead to far sub-optimal results, thus rendering the
  concept infeasible.

- **User convenience**
  From a user perspective, it is essential that a BESS is always capa-

ble of fulfilling its primary purpose. At the same time, system owners are unlikely to be willing to manually interact with the charge
scheduling mechanism to ensure the system is working properly.
In case a BESS is deployed with the primary purpose of buffering
residential PV or powering a PEV, this function therefore needs
to be seamlessly fulfilled without requiring any user interaction.

As a result of these aspects, the deployment of distributed BESSs for
power grid services has remained a niche application which up to now
has rarely passed the stage of pilot testing.

1.2 Approach

This work presents a framework addressing the concerns mentioned in
Section 1.1 by investigating a variety of aspects regarding the optimal
operation of ESSs in smart grids. The primary objectives of this effort
are two-fold:

- **Providing a blueprint for the implementation of improved
  charge scheduling mechanisms for more profitable operation of ESSs**
  The profitability of ESSs depends on the way charging and dispatching decisions are made given a variety of internal and external parameters. This work aims to pinpoint a number of aspects
  which could improve the financial viability of ESS operation if
  integrated into an ESS' charge control mechanism. These shall
  account for the need for optimal autonomous system operation
  without compromising user privacy.

- **Providing a tool for systematically investigating the financial aspects of various types of ESSs in different markets**
  There is a large variety of different types of ESSs including various
  types of battery technologies, flywheels, thermal storage systems
  and supercapacitors. All of these solutions exhibit different technical properties and different cost structures, making them suitable
  for different applications. Fundamental differences cannot only be

observed across different storage technologies but also for different products of the same storage type. The objective therefore is to provide the capability of assessing these technologies with regard to their financial viability based on a highly detailed technical level and to be able to identify conditions under which the deployment of certain technologies is profitable.

Following these objectives and the concerns outlined in Section 1.1, the design decisions of the developed framework and the scheduling methodology are driven by the following presumptions:

- **Investigating the financial viability of ESSs in different markets requires a high degree of versatility**
  Being capable of addressing a large number of different technologies in different markets requires a versatile solution. In order to account for this aspect, the framework is designed with the objective of adaptability by implementing a modular and extensible architecture. This shall allow easy modification, replacement, extension and adding of models to address different types of ESSs and different framework conditions. Code, parameter data and input data are strictly separated in order to allow for easy and systematic parameter variations and investigation of different scenarios.

- **The decision for adopting a certain ESS technology is primarily driven by financial considerations**
  As economic efficiency can be most easily achieved through markets, the implemented scheduling approach is based on market principles, meaning that charging and dispatching decisions should be made based on cost-benefit considerations for the individual system. This shall achieve maximum profitability of the ESS which is presumed as being the ultimate criterion for an investment decision. For this purpose, real-time electricity prices, price forecasts and, for the case of BESSs, battery degradation are considered. This information is utilized to determine optimal charge/dispatch schedules with regard to a certain objective. This objective can

simply be to charge the battery at the lowest possible cost for the application as a PEV or to earn money by providing services to the power grid.

- **The system owner wants to remain in control without any need for manual interaction**
  As outlined above, system owners may be hesitant to share information or to pass control over their system to another entity. The scheduling decision therefore remains entirely on the side of the ESS for maximizing user acceptance. The methodology allows for the consideration of constraints such as a driver's mobility pattern in order to guarantee that the battery is always sufficiently charged when energy is needed for driving.

1.3 Contributions

It is distinguished between methodological contributions as well as contributions in terms of the application of the developed methodology. The methodological contributions of this work presented in Chapters 3 and 4 can be summarized as follows:

- A framework for optimal charge scheduling of grid integrated ESSs is presented. The framework features a modular architecture which shall allow easy modification, exchange and extension of models to account for the large variety of different types of ESSs, system components, parameter values and market conditions. The framework implemented in Python therefore consists of a collection of individual modules representing certain functional parts of the system. Parameters and input data are structured in XML documents and a PostgreSQL database allowing systematic investigations of different scenarios. For computing optimal charge/dispatch schedules for participation in *demand-side management* (DSM) schemes or smart PEV charging, the implemented scheduling mechanism considers real-time electricity prices, price forecasts and battery degradation. The approach also allows the consideration of additional constraints which may result from using an ESS for another purpose such as delivering energy to a residential load or to satisfy a driver's mobility needs in the case of a PEV battery.

- Battery degradation is identified to be an important cost driver when employing batteries for grid services. As this aspect has not received sufficient attention in previous scheduling approaches, a battery degradation monetization model is presented and implemented as part of the framework. The model is formulated in a generic way to serve as a blueprint for integrating customized degradation models matching any particular cell type in use. This is important since different cell types can show very different degradation behaviors so that in general each cell type has to be described by its own model. The implemented degradation model effectively accounts for battery aging during operation and keeps track of the battery's *state of health* (SOH) as the cells age. This includes the simultaneous consideration of calendar and cycle aging with regard to both capacity and power fade. Through monetization of physical degradation, this allows the computation and minimization of battery depreciation costs during operation. Part of the set of battery models are also an *equivalent circuit* (EC) model and a charging model which account for the different charging modes such as *constant current* (CC) and *constant voltage* (CV) charging. Degradation-aware scheduling is not only relevant if the BESS is employed for providing power grid services but can also lead to considerable cost reductions for PEV charging as elaborated in this work. A scheduling strategy minimizing battery degradation is also a means to effectively reduce battery prices since it allows the depreciation of a purchased battery over a longer time horizon.

- The framework integrates with the power system simulation tool CityMoS Power which was collaboratively developed during the preparation of this work and presented in [5, 6]. CityMoS Power can be deployed to investigate the impact of PEV charging onto the power grid so that the combination of both frameworks allows the comparison of uncoordinated and smart charging strategies regarding their implications on power system stability [7]. This facilitates the development and the testing of charging strategies not only in an isolated manner but also with regard to the overall power network.

The contributions in terms of an application of the methodology presented in Chapter 5 are as follows:

- Building on mobility patterns generated using the large-scale traffic simulation platform CityMoS Traffic, the framework is used to investigate the charging needs of Singapore PEV drivers in the case of an electrification of road transportation. Using this simulated data as an input for the scheduling framework, the costs and benefits which result from different charging strategies are presented and put in context with the operating costs of common *internal combustion engine vehicles* (ICEVs). In the greater context of this work, the traffic simulation was also deployed for large-scale optimization of PEV charging infrastructures [8, 9] and for investigating the potential for shared mobility concepts [10].

- The framework is deployed to assess the financial benefits of smart charging mechanisms and the profitability of deploying small-scale BESSs for providing services to the power grid. For this purpose, the implemented scheduling mechanism is thought of as an additional decision layer attached to a *battery management system* (BMS) which makes charging and dispatching decisions in order to maximize its individual utility. Previous economic viability considerations relied on strongly simplified assumptions such as average electricity prices, neglected important aspects such as battery degradation or ignored the optimization potential resulting from exploiting price forecasting information. Using the presented framework, these aspects can be taken into account, thus allowing a more thorough assessment of costs and benefits. By performing computations using price data from various markets, the prof-

itability of energy arbitrage in these markets is investigated and
the sensitivity of profits with regard to various design parameters
is assessed. The results inform about the conditions under which
energy arbitrage using BESSs can be profitable.

1.4 Scope of This Work

The presented work is part of an effort to address the challenges
of transport system electrification, power system stability and PEV
grid integration from a holistic perspective. This has led to the development of the simulation tools CityMoS Traffic and CityMoS Power
with the latter being closely related to the presented scheduling framework. While some contributions were made to CityMoS Traffic and
especially CityMoS Power in the course of the preparation of this
work, this thesis is focused on the scheduling framework while including a few applications where CityMoS Traffic and CityMoS Power are
used.

In accordance with the paradigms underlying the development of
CityMoS Traffic and CityMoS Power, this work pursues a simulation-
based approach aiming for versatile applicability and extensibility.
The framework is developed from scratch and combines various component models and an optimization mechanism. In terms of storage technology, the models and parametrizations are concerned with
small-scale BESSs, and more specifically *lithium-ion* (Li-ion) BESSs,
which is a consequence of the focus on road transport electrification
of the overarching project.

From an application perspective, this work primarily isolates the
financial aspects regarding profits incurring for BESS owners when
participating in real-time electricity markets. This focus is set because of the general controversy of BESS profitability and grounds on
the hypothesis underlying this work that appropriate consideration of
battery degradation can lead to profitable operation. Notable weight
is therefore attributed to the various aspects of battery degradation

on operating costs. The conducted investigations take an isolated view on individual BESSs as price takers so that the possible influence of BESSs on electricity prices as well as concepts for *aggregators* bundling BESSs to *virtual power plants* (VPPs) are beyond the scope of this work. For further reading on this topic it is referred to the literature on aggregation concepts for PEVs in [11–18], for smart loads in [19, 20] and on coalition formation for VPP aggregators in [21–24].

1.5 Organization of This Work

This thesis is organized into a part describing the relevant fundamentals related to this work, a part describing the methodology and its implementation as well as into a part where the framework is applied to a number of different scenarios. An overview of the content of the individual chapters is given as follows:

The Big Picture (Chapter 2) In a broad sense, this work deals with the grid integration of ESSs in general and with PEVs in particular. In this chapter, the relevant fundamentals related to this topic are established. This comprises a brief introduction into some aspects of smart grids with a focus on DSM and the relevant control paradigms. As this work focuses on economic feasibility considerations of ESSs, an important aspect is presented by electricity markets where energy and capacity are traded. This chapter therefore also provides an overview of the various types of markets, the services traded at these markets and their corresponding pricing schemes. Furthermore, an introduction into BESSs is given with a particular focus on their suitability for the various market types and the development of battery prices which is particularly relevant for the economic feasibility considerations. The chapter finally highlights the relevant aspects related to grid integration of PEVs. This comprises a simulation study investigating the power grid impact of PEV charging conducted in the context of this work which showcases the need for smart charging strategies.

Framework Architecture (Chapter 3) This chapter outlines the purpose of the developed framework and provides a detailed description of its architecture and components. As the framework is used in conjunction with the traffic simulation CityMoS Traffic and the power system simulation CityMoS Power, both of these frameworks are also briefly described in this chapter.

Scheduling Approach (Chapter 4) In this chapter, the scheduling approach implemented as part of the framework is described in detail. This comprises a formulation of the optimization problem addressed by the implementation and a description of the solving method. Furthermore, the models developed to handle the problem including a battery model, a charging model and a battery degradation monetization model are presented in detail.

Applications (Chapter 5) This chapter presents the application of the developed framework to a number of different scenarios. This shall on the one hand provide an understanding of the functioning of the presented methodology, and address the questions on the financial aspects of smart charging and BESSs on the other one. The application provides an insight into the relevance of appropriate modeling and monetization of battery degradation as part of a methodology for optimal charge scheduling. The framework is also deployed for comparing different PEV charging strategies to demonstrate the financial implications smart charging and the provision of services to the power system can have for a PEV owner. Finally, a study investigating the financial benefits of energy arbitrage using BESSs in various markets is presented. While a definite conclusion on the economic feasibility of certain applications depends on a variety of parameters, this study gives an insight into the relevant aspects determining profitable operation.

Conclusion (Chapter 6) The final chapter summarizes the presented content and findings and provides an outlook on possible future work.

2 The Big Picture: Smart Grids, Electricity Markets, Energy Storage Systems and Electric Vehicles

2.1 Introduction

In this chapter, the fundamental concepts and terms related to the presented work are introduced. Section 2.2 gives a brief overview of a few smart grid related aspects. Due to the wide range of topics which are subsumed under the term *smart grid*, the focus is set on the fundamentals related to DSM which are most relevant in the context of this work. For integrating DSM concepts into power systems, they have to be seen in the context of electricity markets. Section 2.3 therefore introduces the fundamental concepts and types of electricity markets which most advanced power systems have in common. While there is a variety of different technologies which can be employed as smart loads for DSM, the technology in the focus of this work is *distributed energy storage* (DES). Section 2.4 therefore provides a brief introduction into energy storage concepts with a focus on batteries. One type of DES is given by PEVs which can in principle be considered for DSM services. Section 2.5 therefore discusses the challenges and opportunities related to PEVs in this context. In Section 2.6, an overview of techniques for using mobile and stationary BESSs to provide services to the power system is given. Section 2.7 summarizes the main conclusions from this chapter which are the foundation for the concept developed in this work.

© Springer Fachmedien Wiesbaden GmbH, part of Springer Nature 2019
D. Pelzer, *A Modular Framework for Optimizing Grid Integration of Mobile and Stationary Energy Storage in Smart Grids*, https://doi.org/10.1007/978-3-658-27024-7_2

2.2 Smart Grids

The term *smart grid* subsumes a wide range of enabling technologies, *information and communications technology* (ICT) infrastructure, hardware and practices which aim to make the power distribution infrastructure more reliable, secure and efficient and facilitate
the integration of intermittent RES [25, 26]. Most importantly, the
development of a smart grid marks a transition from a centralized
system based on large power generators on one side and consumers
on the other one towards a decentralized architecture. In the latter,
consumers may also be considered *prosumers* which do not only act as
loads but may also be energy producers themselves [27]. One important aspect of a decentralized system is the intelligent management of
electric loads and the temporary storage of energy required to ensure
stable and efficient operation of such decentralized systems. This is
the overarching topic of the presented work which is why this section
introduces some basic concepts regarding this issue. For a comprehensive technical overview of smart grids, the reader is referred to
the exhaustive elaborations in [28, 29]. A collection of contributions
on the role of PEVs in smart grids can be found in [30, 31].

2.2.1 Demand-Side Management and Energy Storage Systems

The current power system infrastructure functions to a large extent as
a just-in-time inventory system in which energy needs to be generated
precisely at the time it is needed. Since demand-supply imbalances
immediately manifest themselves in frequency fluctuations threaten-
ing system stability, generation capacities need to be continuously
ramped up and down by exactly following load requirements. This increases equipment wear, reduces power plant efficiency and therefore
has an undesirable effect on fuel consumption and emissions. Furthermore, the system needs to be sized to satisfy worst case demand
peaks which results in considerable over-capacities. Finally, with the
increasing shares of RES in power generation, the system loses flexibility on the supply side because the output of most RES is hard to

adjust. Tackling these challenges is one of the key tasks of a smarter grid.

In principle, there are two approaches to address these issues. The first one is to build on *smart loads* which aim to perform the adaptation on the demand side instead of the supply side. In this case, intelligent loads adapt their power consumption to the current state of the system, reducing power consumption when system load exceeds generation and vice versa. The second approach is to decouple generation and consumption by buffering excess energy in ESSs and feeding this energy back into the grid in the case of power shortages. This could essentially eliminate the distinction between peak and base load generation which would allow loads to be served by the lowest cost energy resources at any time [32]. Concepts where the balancing is performed on the demand side are generally termed DSM, *demand-side response* (DSR) or *demand-response* (DR) [33–35].

As illustrated in Figure 2.1, various load shaping objectives can be accomplished by DSM measures including peak shaving, valley filling, load shifting, strategic conservation, strategic growth and flexible load shape [36]. Among those, the forms of load management relevant on short time scales include peak shaving, valley filling and load shifting, all of which aim to smoothen the demand curve throughout the course of the day. Strategic conservation, strategic load growth and flexible load shape are practices addressing greater time horizons, e.g., through incentivizing or disincentivizing the sales of products of which the use has the desired effect on the load curve. Apart from these high-level goals altering the demand curve, lower level goals addressing shorter time frames or more localized system components can be pursued. These include aspects such as frequency regulation, reactive power compensation, managing network congestion or avoiding transformer overload [33, 34].

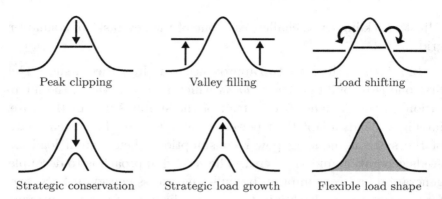

Figure 2.1: Basic load shaping techniques adopted from [36].

2.2.2 Control Paradigms

Achieving the objectives underlying DSM generally requires some sort of control entity as well as one or multiple control objects which have the capability of adapting their power demand in response to a control signal. As outlined in the following, in this regard three basic control paradigms can be distinguished:

- **Centralized control**

  In a *centralized control* or *direct load control* setting, a single entity makes the control decision by sending control signals to a load or generator. The control object is typically offered a compensation to provide an incentive for participating in such control setting [35]. From a system perspective, this is a relatively simple approach because all information required is available at one point and the controller is able to foresee the effects of the control decision. This allows an effective improvement of network capacity utilization. It does, however, limit the autonomy of the controlled object. Furthermore, the controller requires a potentially large amount of information on each individual control object which is a challenge from an information processing perspective and a concern from a privacy point of view. This is certainly an issue with regard to PEVs because a driver may, for instance, have objections against disclosing his mobility behavior and may feel discomfort with another entity controlling the SOC of his battery. Furthermore, the controller is unlikely to be capable of effectively considering vehicle specific aspects such as battery degradation since the degradation behavior of the particular battery types is not known to the controller. As different operating conditions cause different battery strain, the controller may therefore contribute to premature battery degradation, thus imposing financial harm on the owner of the battery. Another drawback of centralized control is the growing complexity with the number of controlled objects which makes the problem of computing optimal charge/discharge schedules increasingly difficult to solve [37].

- **Distributed control**
  Distributed control, decentralized control or price control is a form of indirect control where the controller sends price signals to the control objects. Based on their individual utility functions, these objects then respond accordingly. Price control is therefore simpler than centralized control from a communication and information perspective because the controller needs less information from the control objects and only uni-directional communication is required. Since the controller does not have to optimize the behavior of each individual control object, this approach also allows for easier scalability. The setting can also be considered to have a higher fault tolerance since there is not only one controller as a single point of failure. As the control remains on the user side, higher user acceptance can also be expected. Furthermore, the individual user may be able to better optimize the scheduling since technology-dependent factors such as battery degradation can be accounted for in an easier way. From a design perspective, however, the control problem is more difficult because the exact response of the system is unknown to the controller. This may result in avalanche effects or system oscillations if a large amount of loads or generators respond in a similar way [38, 39]. Appropriate performance

therefore requires the ability to predict or forecast the response of the system thus making this setting more challenging than centralized control.

- **Transactive control**
  The *transactive control* paradigm is a form of market-based mechanism where system- and component-level decisions are based on economic contracts negotiated between the system's components. This is typically an iterative process based on bi-directional communication where the controller and the control object attempt to reach an equilibrium. With regard to large-scale energy infrastructures, this concept has recently been explored in a variety of projects [40–42] and is subject to growing research efforts.

More a system design feature rather than a control paradigm is the concept of *hierarchical control* [43–45]. In this case, different hierarchy levels may implement any of the different control paradigms which is a measure to reduce the control problem's complexity.

2.3 Electricity Markets

Power systems implement different markets for different kinds of services. This section gives a brief overview of the fundamentals of electricity markets and introduces relevant terms which will be referred to later in this work. A basic understanding of the different market types and services is essential in order to understand which of them are suitable for BESSs and what constraints need to be taken into consideration.

While the implementation details of individual electricity markets greatly vary between regions and countries, some of the basic concepts are quite universal. The following sections therefore focus on these generally applicable principles. For more detailed information, the reader is referred to the reviews in [46–48] and for particular markets to the market specifications provided by the individual market operators.

2.3.1 Market Operators

An electric power system, transmission lines and the corresponding wholesale electricity market are typically administered by organizations which may be owned by a transmission grid company or which may be fully independent and non-profit oriented. In the U.S., these are generally termed *independent system operators* (ISOs) or *regional transmission organizations* (RTOs) with the European equivalent being called *transmission system operators* (TSOs). At a smaller scale, there may further be *distribution system operators* (DSOs) which oversee the distribution system. As long as no distinction is required, these entities will be referred to as PSOs in the remainder of this work. These operators may underlie overarching entities such as the *Federal Energy Regulatory Commission* (FERC) in the U.S. or the *Union for the Coordination of the Transmission of Electricity* (UCTE) in Europe which elaborate standards and recommendations for system operation.

Agents participating in the wholesale markets administered by the PSOs are generators on one side and resellers such as electric utility companies, competitive power providers and electricity marketers on the other one. The underlying principle of any of the wholesale markets is to match offers from generators and bids from buyers to achieve an equilibrium price which can be referred to as the *market clearing price* (MCP). The last stage of the electricity supply process is implemented through retail markets where resellers serve the end-users.

2.3.2 Market Types and Service Classes

Wholesale markets can be categorized according to the time scales they are operating on and the services they are dealing with. The short term *market types* which are most relevant in the context of ESSs cover the following three groups [46]:

- Day-ahead market which allocates resources for the following day

- Intraday market for settling hour-ahead supply

- Real-time market for intra-hour system balancing

These markets are typically spot markets which involve immediate settlement. In addition, there are usually markets addressing longer time frames in the order of weeks to years. As opposed to the short term markets which are generally implemented as exchanges, these are more typically *over-the-counter* (OTC) markets which may come in the forms of both futures markets as well as spot markets. The framework presented in Chapter 3 uses this categorization to specify which market the ESS is participating in.

The commodities traded in these markets are energy as well as power capacity for regulation and reserve. As outlined in the following, in this regard three different categories of *service classes* can be distinguished:

- **Generation**
  Generation for base load is typically performed by power plants producing energy at a low cost but without the capability for fast power output adaptation. Peak loads, in contrast, are handled by more flexible generators which in turn exhibit higher generation costs. While generation for peak demand may be performed by large-scale energy storage facilities such as hydro-electric power plants, BESSs are too costly to be suitable for any of these services.

- **Load smoothing**
  Load smoothing including peak shaving, valley filling and load shifting addresses variations of the load curve. Load smoothing can technically be achieved by flexible consumers which adapt their power input to the load curve as well as by BESSs.

- **Ancillary services**
  Ancillary services ensure power system stability by means of frequency control. In this context, different service sub-classes can be distinguished which differ by their response times. According to the FERC, in North America they are categorized into the classes

regulation reserve and *operating reserve*, the latter of which consists of the subcategories *spinning reserve, non-spinning reserve and replacement reserve*. Regulation reserve is the service with the fastest response time, shortest service availability and the highest prices. For regulation, it is distinguished between *up-regulation* in case power is generated to counteract a frequency drop and *down-regulation* in the opposite case. The spinning reserve consists of synchronized generation capacities which are capable of responding to a contingency within a time frame of minutes. Non-spinning reserves are idle capacities which can be ramped up in a short period of time, typically within around 30–60 min. The slowest service is the replacement reserve which consists of low cost generation capacities of which the purpose is to substitute the faster and more expensive reserves to reduce costs.

The European equivalent defined by the UCTE consists of the three sub-classes *primary reserve, secondary reserve* and *tertiary reserve* (also termed *minute reserve*). Primary reserve is the service with the fastest response which has to be deployed within 15–30 s after a frequency deviation and which needs to be provided for a duration of up to 15 min. It is supplied mainly by conventional power plants. The secondary reserve replaces primary control until frequency has reached its nominal level. It is available in a time frame of seconds up to 15 min after an incident. Tertiary reserve is manually activated and typically lasts for 15–20 min, sometimes for extended periods of up to 2 h. Its primary purpose is to release primary and secondary reserves for other contingencies.

The framework presented in Chapter 3 uses this categorization to specify the service class of the market the BESS is participating in.

2.3.3 Pricing Schemes

With regard to types of payments and pricing schemes, different categories can be distinguished. These comprise the type of commodity which is traded as well as the geographical and temporal resolution of the pricing scheme.

2.3.3.1 Differentiation by Commodity

The commodities traded at electricity markets can be categorized into energy and capacity. Energy payments apply to the units of energy which are dispatched and are therefore paid in units of Wh (or more practically MWh). In contrast, capacity payments are made for keeping power generation capacity available for a certain period of time in case the demand exceeds current supply. These are remunerated in units of W·h (or more practically in units of MW·h). The framework presented in Chapter 3 uses this categorization to specify the payment type the optimization leverages on.

2.3.3.2 Differentiation by Geographical Location

In terms of geography, three different pricing schemes can be distinguished which determine the region in which a price applies:

- **Nodal marginal pricing**
  With *nodal marginal pricing* (NMP) or *locational marginal pricing* (LMP), prices differ between the different nodes of the system. The prices themselves are typically based on an *optimal power flow* (OPF) implementation. These prices reflect the incremental cost of supplying the next unit of demand at a particular node (*shadow price*). They account for locational patterns of generation and load as well as for the local limitations of the transmission and distribution system by considering *marginal energy cost* (MEC), *marginal loss cost* (MLC) and *marginal congestion cost* (MCC).

- **Zonal marginal pricing**
  With *zonal marginal pricing* (ZMP), prices are uniform within a

particular zone which is a defined geographical area. Zonal prices are typically computed as weighted averages of nodal prices.

- **Uniform marginal pricing**
  With *uniform marginal pricing* (UMP), there is one uniform price for the entire control area. Again, this price typically results from a weighted average over the lower level prices.

The framework presented in Chapter 3 uses this categorization to specify the maximum spatial resolution for schedule optimization.

2.3.3.3 Differentiation by Temporal Resolution

With regard to temporal resolution of pricing, three main categories can be distinguished:

- **Fixed pricing**
  Fixed pricing is still very common in retail markets. While fixed pricing is easy to implement, it does not provide any incentives for consumers to adapt their consumption in favor of smoother load curves. This leads to an inefficient outcome because it generally requires power generators to be ramped up and down more often, thus resulting in higher generation costs.

- **Time-of-use pricing**
  Time-of-use (TOU) pricing takes the concept of fixed pricing one step further towards more efficient use of electricity by distinguishing prices depending on the time the electricity is used. This is typically realized in terms of peak and off-peak prices which shall shift demand towards periods where overall system load is low. A drawback of TOU pricing, however, is that it may trigger demand peaks right after the high-price period has elapsed [49].

- **Critical peak pricing**
  Critical peak pricing (CPP) is a form of TOU pricing. In contrast to TOU prices, CPP may affect prices only during very few events a year when peak load is extraordinarily high. Just as TOU pricing,

CPP may come with the drawback of high demand as soon as prices turn back to normal.

- **Real-time pricing**
  Real-time pricing is the most desirable concept from the perspective of efficient resource allocation. In this case, prices are set for certain periods of time in the range of an hour or less. While real-time prices are still not common in retail markets, they would allow consumers to adapt their demand in accordance to the price signals they receive. In many wholesale markets, price forecasts are available or can be generated using time series analysis methods so that a consumer could elaborate schedules maximizing its individual utility.

The framework presented in Chapter 3 uses this categorization to specify the maximum temporal resolution for schedule optimization.

2.4 Energy Storage

As elaborated in Section 2.2, ESSs may play an increasingly important role in a future smart grid. There are a number of different forms in which energy can be stored including i) chemically and electro-chemically (e.g., batteries, hydrogen, fossil fuels, bio fuels), ii) electrically (e.g., capacitors, superconducting magnetic energy storage), iii) mechanically (e.g., compressed air, hydro energy storage, flywheels) and iv) thermally (hot water/vapor, latent heat). With regard to PEV applications and small-scale ESSs, a dominant role is played by batteries which are therefore of primary interest in this work. This section presents a few relevant aspects including the price development of batteries and discusses their suitability for various power grid services. The electro-chemical basics of battery operation are not of primary importance to this work; for further details in this regard, the reader is therefore referred to [32, 50–52].

2.4.1 Battery Technologies

There is a variety of battery technologies available on the market including *lead-acid* (Pb-Ac), *nickel-cadmium* (Ni-Cd), *nickel-metal hydride* (Ni-MH), *sodium-sulfur* (Na-S) and Li-ion batteries. With regard to PEVs and to an increasing degree for stationary BESS applications, Li-ion type batteries have emerged as the technology of choice [50]. For mobile and small-scale applications, Li-ion batteries are particularly suitable as the high negative potential of lithium at low atomic weight leads to high voltages at low weight, thus implying high energy densities and a high power density [53]. During operation, Li-ion battery electrodes furthermore undergo little structural changes which results in a long cycle life of >1 000 cycles at 80 % *depth-of-discharge* (DOD) [53]. Among secondary battery technologies, Li-ion batteries also show the best charge retention with self discharge rates of about 2 % per month at ambient temperatures [54]. Furthermore, Li-ion batteries have comparably high potential for fast charging and are considered to exhibit no memory effect [55].

Li-ion type batteries can be categorized into a number of technologies including *lithium-nickel-cobalt-aluminum* (NCA), *lithium-nickel-manganese-cobalt* (NMC), *lithium-manganese oxide* (LMO), *lithium-titanate oxide* (LTO) and *lithium-iron phosphate* (LFP) with different characteristics in terms of energy density, power density, performance, safety, life span and cost [56].

One common downside of the various Li-ion type batteries compared to alternative technologies has been their high price. For this reason, employing Li-ion batteries for grid applications and even for PEVs has been uneconomical in the past. Due to increased research attention and economies of scale, however, prices have seen a steep decline in recent years. In a comprehensive study in [57], it is shown that for the industry as a whole, average prices of Li-ion batteries for PEVs declined by 14 % annually between 2007 and 2014 from above

$1 000 per kWh[1] to $410 per kWh. For leading manufacturers, prices by 2014 are assumed to be even lower at around $300 per kWh. For the end of 2016, a recent study [58] estimates battery pack prices at $230 per kWh which continues the trend outlined in [57]. The same study expects PEV battery pack prices to fall below $190 per kWh by the end of the decade and to drop below $100 per kWh by 2030. In an even more recent study, it is argued that previous forecasts systematically underestimated the decrease of PEV battery prices [59]. The authors therefore developed a two-factor learning curve model considering economies of scale and innovation activity represented by cumulative international *Patent Cooperation Treaty* patents which is shown to better fit price trajectories than previous studies. An optimistic forecast based on this model predicts PEV battery pack prices of $160 per kWh for 2018, a considerably faster decrease than alternative estimates. This study argues that the forecasted price development could rapidly turn BESSs from a niche market for protection against blackouts to a widely deployed technology. This underlines the validity of the efforts of individual industry projects such as the Tesla Motors Gigafactory which aims for an output of 35 GWh in 2020, corresponding to the global production volume of the year 2013 [60].

An aspect which indirectly influences battery operating costs is the battery's lifetime. Due to their intrinsic complexity, the study of aging mechanisms of Li-ion batteries has developed at a rather slow pace. Thanks to the improvement of experimental and theoretical characterization tools and the need for more durable technologies, the research efforts in this direction have, however, recently gained traction [50]. Apart from technological improvements, scheduling strategies which better account for battery degradation processes as presented in this work can therefore make another contribution to effective cost reductions.

---

[1] Unless stated otherwise, monetary numbers are given in US$.

2.4.2 Suitability of Battery Energy Storage Systems for the Different Services and Markets

Batteries have a number of properties which make them attractive for application for DSM programs. They are compact in size and are therefore ideal for distributed storage. This is particularly useful in the context of RES integration since the energy can be stored where it is generated so that no additional grid load and transmission costs incur. BESSs are modular so that systems can be easily scaled in case requirements change. They further have a very short response time so that they are able to almost immediately respond to a control signal requesting them to charge or discharge. Finally, batteries have a high charge/discharge efficiency thus keeping energy losses low.

As BESSs can provide energy only for a limited period of time and at a comparably high price, they neither qualify for base load nor peak load generation. They do, however, technically qualify very well for the participation in DSM programs since they can consume or provide energy at relatively high powers as compared to common household consumers. Due to their capability to almost instantly respond to a control signal, they are further suitable for providing ancillary services. This particularly applies to those service where a fast response is required and where dispatch times are short. In accordance with the discussion in Section 2.3.2, these characteristics are best fulfilled by frequency regulation/primary reserve. For these services, prices are relatively high and dispatch quantities are small compared to the procured capacity. Furthermore, regulation up and regulation down are typically balanced so that the overall state of charge of the device is not considerably affected. While spinning reserve/secondary reserve is also considered as an application for batteries albeit lower prices, non-spinning reserve/tertiary reserve is considered unsuitable because required dispatch times are high, therefore leading to battery depletion and deep cycles resulting in accelerated degradation.

2.4.3 Drivers for the Market Development of Battery Energy Storage Systems

Acknowledging the trend towards small-scale ESSs, PSOs and regulators have started establishing framework conditions aiming to facilitate their integration into the power system. Examples include the FERC Orders 745, 755 and 784 which require RTOs to treat DSM aggregators the same way they treat conventional power plants and which open the grid for storage systems by favoring BESSs for ancillary services. Another example is a recently released FERC rule (Docket No. RM16-23-000) [61] demanding RTOs and ISOs to revise their tariffs in order to accommodate the participation of DER in wholesale electricity markets. Similarly, the TSOs in Germany published specific requirements clarifying the framework conditions for the participation of BESSs in ancillary service markets [62, 63].

Revised regulations regarding feed-in tariffs for RES may turn out to be an additional factor driving the growth of residential BESS installations. As an example, the 2012 and 2014 revisions of the *German Renewable Energy Sources Act* (German: EEG) led to significant reductions of the feed-in tariff combined with feed-in quantity limitations for owners of PV systems. This forces PV system owners to consume larger shares of PV-generated energy themselves. As the output of a PV module can, however, not be controlled, this can be most effectively achieved by installing a local BESS which buffers excess energy. The installation of BESSs is incentivized through subsidies by the German KfW bank.

Apart from the development of regulatory frameworks, there have also been projects to investigate and showcase the integration of small-scale producers into the power system. An example for this is the EcoGrid EU initiative, a large-scale demonstration project regarding the integration of small-scale producers into real-time markets [46].

A more comprehensive *political, economic, social and technological* (PEST) analysis investigating the different aspects related to BESSs was carried out in the context of this work and is presented in [64].

2.5 Plug-In Electric Vehicles

With the growing adoption of PEVs as alternatives to ICEVs, PEVs are emerging as increasingly important actors in the power system. As loads with a high power consumption and a relatively large storage capacity, their deployment can have both detrimental and positive effects on power system infrastructures. This section provides a brief overview of the market aspects of PEVs in order to demonstrate their relevance, their possible power system impact in case they are operated as inflexible loads as well as on their potential for participating in DSM services. As an illustration, this section contains a simulation study investigating the impact of PEV charging on the Singapore power network which was prepared in the context of this work and published in [5].

2.5.1 Market Aspects

Global sales of *battery electric vehicles* (BEVs) and *plug-in hybrid electric vehicles* (PHEVs) have grown from approx. 6 000 units in 2010 to 750 000 units in 2016, resulting in more than 2 million PEVs on the roads by the end of 2016 [65]. While the market development of PEVs has significantly fallen below the initial development targets such as the 1 million PEV goal by 2020 in Germany, the recent development indicates an accelerated market evolution. In the coming years, this can be expected to be driven further by the rapid decrease of battery prices discussed in Section 2.4. Different studies arrive at the conclusion that PEVs become economically competitive with ICEVs on a TCO basis at battery pack prices between $ 125 per kWh and $ 300 per kWh, depending on fuel prices [57, 66–68]. Taking into account the expected decrease of battery prices along with these considerations on economic parity with ICEVs and assuming a further development of charging infrastructures, PEVs may therefore soon develop from a niche to a mass market product. This underlines the importance of addressing the challenges of PEV grid integration to avoid future bottlenecks in power infrastructures.

2.5.2 Power Grid Impact

In the power grid, undesirable effects resulting from uncoordinated charging can in principle occur in any of the different voltage levels, namely the *high voltage* (HV), *medium voltage* (MV) and the *low voltage* (LV) grid. Bottlenecks can manifest themselves by exceeding the capacity of power generation infrastructure or by exceeding the maximum power rating of substations or power lines. The power grid impact of PEVs depends greatly upon the way they are operated, as outlined in the following.

2.5.2.1 Plug-In Electric Vehicles as Inflexible Loads

The literature identifies various challenges regarding the large-scale integration of PEVs into distribution systems. Uncontrolled charging is expected to increase peak demand with implications for generation facilities and transmission systems as outlined in studies for the U.S. [69–71], Germany [72] and a number of other countries [5, 73]. It, however, is acknowledged that while effects on generation and transmission system level cannot be ruled out entirely, the main effects will be seen in distribution systems [74–76]. These can include increased power losses [77], overloading of transformers [44, 78] and power lines [79], as well as voltage drops [80, 81]. These effects may particularly affect the LV and to a potentially lesser degree the MV [44] level. Further reviews on the power grid impact of PEV charging can be found in [82–85].

While definite conclusions on the power grid impact of PEVs depend on a large number of parameters and may only be valid for a particular power network, this indicates that measures to ensure stable power grid operation even in the case of large PEV populations are absolutely essential.

2.5.2.2 Plug-In Electric Vehicles as Flexible Loads

While incremental infrastructure investments can enable power systems to accommodate increasing numbers of PEVs, a cheaper alternative can be to treat PEVs as flexible loads which can prevent contingencies and even improve power system stability [7]. In [86] it is shown that smart charging and discharging during peak hours could reduce the need for incremental distribution grid investments to 30–40 % as compared to an uncoordinated charging scenario where charging simply starts as soon as a power connection is established. The authors in [87] demonstrate that local prices in accordance with the current loading of the transformer can reduce load peaks by more than 80 % if PEVs are treated as price-responsive actors. Another price-responsive approach is presented in [88] which also indicates that local load peaks can be reduced if PEVs are sensitive to charging prices. These examples show that while PEVs as inflexible loads may pose risks to power system stability, smart charging approaches are suitable to effectively mitigate the power grid impact resulting from PEV charging.

2.5.2.3 Simulation Study on the Example of Singapore

The power grid impact of PEV charging was also investigated in the course of this work using a simulation-based approach on the example of Singapore [5]. For this purpose, the tempo-spatial power demand was simulated employing the traffic simulation framework CityMoS Traffic[2] which is briefly introduced in Section 3.5.1. Using

---

[2] Formerly denoted SEMSim Traffic.

Table 2.1: Properties of the Singapore PNM.

Property	LV[a]	MV[b]	HV[c]
Branch length, average [m]	31	210	5 280
Branch length, total [km]	3 789	4 198	491
Characteristic path length [m]	20	145	4 520
Clustering coefficient, average	0.05	0.00	0.08
Mean degree	1.98	2.13	4.33
# Nodes	123 403	18 725	43
# Edges	121 971	19 955	93
# Substations, supplying	11 726	824	10
[a] LV: 0.4 kV [b] MV: (6.6, 22, 66) kV [c] HV: (230, 400) kV			

the power system simulation framework CityMoS Power[3] as introduced in Section 3.5.2, the resulting load flow in the power network
was simulated. Since sufficiently detailed information on the Singapore power network is not publicly accessible, a methodology for the
bottom-up synthesis of power networks based on tempo-spatially resolved demand data was developed in the context of this work and
published in [6]. Using this approach generally termed *power system planning* (PSP) [89], a *power network model* (PNM) serving as
a proxy for the Singapore power system was generated. For this
purpose, the number [90], location [91] and power demand [92] of
117 852 geographically distributed consumers connected to the various voltage levels with a total peak power demand of 6.34 GW[4] was
taken as an input parameter. Additionally, publicly accessible information on substations and power plants [93] as well as on power
lines [94] was used. The methodology developed for this purpose is
briefly described in Section 3.5.2.1. The network specifications of the
resulting PNM used for the actual power flow simulations are listed
in Table 2.1.

---

[3] Formerly denoted SEMSim Power.
[4] Peak power demand on 12th January 2014.

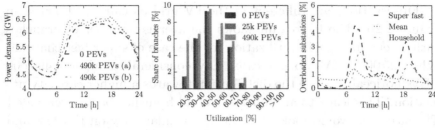

(a) Load curves for differ- **(b)** Share of branches ac- **(c)** Number of substations
ent charging scenarios cording to LV grid with voltage drops
with (a) super fast and capacity utilization. below the accepted
(b) mean charging. threshold for 490 000
PEVs.

Figure 2.2: Simulation results showing the impact of PEV charging on the
Singapore power grid.

The study was conducted for different scenarios with 0, 25 000 and
490 000 PEVs and three simple charging strategies, namely *house-*
hold charging (3.6 kW at residential locations only), *super fast charg-*
ing (120 kW at every location) and *mean charging* (charging at every
location by distributing the charging power evenly over the entire
parking time).

Figure 2.2a shows the influence of different PEV numbers for the
super fast charging and the mean charging strategy on the daily load
curve. It can be seen that 490 000 PEVs exert a measurable influence on the overall load. The temporal demand distribution varies
between both charging strategies with sharper peaks in the super fast
charging case and a higher share at night times for the mean charging
strategy. With an increase of less than 4 % and compared to the overall generation capacity of 12.5 GW [93], the impact of 490 000 PEVs
on the load curve can, however, be considered insignificant. The load
curve for 25 000 PEVs is almost identical to the curve belonging to
the basic scenario which is why it is not shown in Figure 2.2a. It can
therefore be concluded that on an aggregated level no major supply
bottlenecks need to be expected.

Observations are different in the LV grid as can be observed in Figure 2.2b. The histogram shows the average number of branches with a certain capacity utilization for the case of super fast charging. While 25 000 PEVs do not cause any considerable impact, 490 000 PEVs noticeably move the distribution towards a higher capacity utilization. This leads to around 1.7 % of all branches being overloaded. The situation would look different in a similar view on the MV and HV grid which would not show any branches operating above their nominal capacity limits.

This effect becomes even more obvious when looking at Figure 2.2c showing the share of substations in the LV grid which are not able to keep the voltage magnitude at their consumers above 0.95 pu. During peak hours, this affects about 5 % (super fast charging), 2.5 % (mean charging) and 1.5 % (household charging) of all LV substations. In contrast, a similar plot for the scenario with 25 000 PEVs would not show any significant increase in system load.

These results indicate that the HV and MV grid can be expected to remain largely unaffected even if the entire fleet of private vehicles in Singapore would be electrified. In contrast, the LV layer may be negatively affected by uncoordinated charging of large numbers of PEVs resulting in grid congestion and voltage drops. The comparison of various charging strategies also indicates that even large numbers of PEVs could be integrated into the power grid if appropriate measures for smart charging were taken.

This outcome is well in accordance with the related work discussed above. It, however, needs to be noted that these investigations can only provide a rough estimate because the power network can only be considered a crude approximation of the real-world power system. This is because the PSP approach does not consider the temporary evolution of power systems but instead generates a quasi-optimal system from scratch. Furthermore, geographic or environmental constraints are not taken into consideration when placing substations or laying power lines. These may therefore be placed at locations which would not be chosen in reality. Finally, within each voltage level homogeneous specifications for substations and power lines are assumed. This does not reflect the fact that different substations may have different power ratings or that power lines of different qualities may be used. More accurate quantitative conclusions could therefore be drawn with access to greater amounts of real-world data. This should particularly include exact information on actually laid power lines and their over-capacities in the distribution network. Also more detailed tempo-spatial information on the existing power demand would be beneficial for correctly quantifying the additional local loads resulting from PEV charging. A further elaboration on these issues can be found in the context of the discussion of CityMoS Power in Chapter 3.

2.6 Grid Integration of Battery Energy Storage Systems and Plug-In Electric Vehicles

With the growing number of small-scale residential BESSs and PEVs, there are new types of consumers and potential temporary generators being integrated into the power system. As outlined in Section 2.5.2, depending on the way these devices are operated, their deployment can have beneficial or detrimental effects on the system's stability and efficiency. This section discusses a number of aspects regarding the grid integration of stationary and mobile BESSs. The focus of these considerations is on the different ways coordinated operation can be performed as well as on the economic aspects of BESS operation. Further overviews of the role of BESSs and PEVs in smart grids can be found in [38, 74, 84, 95–100].

2.6.1 Classification of Coordination Approaches

BESSs and electrical loads in a power network can be controlled in different ways. As discussed in the following, coordination approaches can be classified according to two main criteria which are the type of coordination and the scope of coordination.

2.6.1.1 Types of Coordination

PEVs and BESSs can generally be operated in an *uncoordinated* (i.e., dumb charging) and a *coordinated* (i.e., smart charging/discharging) way. As discussed in the context of DSM in Section 2.2.1, coordination can be achieved through three main control architectures, namely centralized, decentralized and transactive control. Furthermore, in hierarchical control settings these control paradigms might occur in combination at different hierarchy levels. It can further be distinguished between *uni-directional* and *bi-directional* power flows. While a uni-directional power flow implements charging, bi-directional power flows also allow the BESS to act as a temporary energy source. In the context of PEVs, this concept is termed *vehicle-to-grid* (V2G) [101]. As this work deals with both mobile and stationary BESSs, the more general term *battery-to-grid* (B2G) is used unless the considerations are explicitly limited to PEVs.

An advantage of uni-directional B2G where batteries are treated as smart loads [102] compared to bi-directional energy flows is supposed to be that it does not incur any additional battery degradation [103]. Furthermore, no investments into B2G equipment would be required in this case. As will be shown in Chapter 5, however, the argument of no additional battery wear is only true under certain simplifying assumptions. As an additional downside, potential revenue streams from deploying the battery as an energy source also remain untapped [104]. Despite being arguably easier to implement, uni-directional B2G therefore does not exploit the full optimization potential.

2.6.1.2 Scope of Coordination

In the case of coordinated operation, different scopes or scales of coordination can be distinguished. In the simplest cases, coordination is implemented at a very local level such as a single smart home, a building or a car park. At a higher level, it may be realized at the scale of a microgrid or may also cover a greater area over different LV networks. In a scenario in which the BESS provides services to the power grid, the system integration is typically facilitated by an additional entity which acts as an intermediary between the individual BESS and the electricity market or PSO. This so-called aggregator bundles a potentially large number of BESSs to a single VPP [11, 12]. This is often necessary since actively participating in electricity markets, e.g., by taking part in auctions, usually requires a certain minimal capacity of the participating agent. This capacity often is in the megawatt range, for instance with a minimum capacity of 1 MW in the Singapore market or 10 MW per bid in the Nord Pool regulation market. This requires combining a considerable number of small-scale BESSs to an aggregated entity.

2.6.2 Profitability of Battery Energy Storage Systems for Power Grid Services

By participating in energy markets, owners of BESSs have the possibility of generating profits by providing services to the power system. This can be achieved through energy arbitrage, mostly on intraday price fluctuations, as well as through participating in ancillary service markets. While the technical viability of these concepts has been demonstrated in a great variety of studies particularly in the context of V2G [49, 77, 105, 106], notable doubts regarding their economic profitability remain [3, 95, 98, 100, 107–111].

2.6.2.1 Energy Arbitrage Through Load Leveling

Energy arbitrage through buying during periods of low prices and selling at high-price periods can be one possible revenue stream for BESSs. A number of studies examined the profitability of energy arbitrage through B2G. An investigation using hourly LMP data for the cities Boston, Rochester NY and Philadelphia is presented in [106]. In this work, the authors arrive at the conclusion that annual profits of $ 10–120 can be achieved which cannot be considered enough to justify transaction costs. Battery degradation is identified as the main cost driver which compensates a large fraction of the revenues. A weakness of this study, however, is that battery degradation costs are statically fixed at $ 0.042 per kWh. It will be shown in Chapter 5 that this is a very strong assumption with considerable implications on the validity of the conclusion. The relevance of battery degradation as a limiting factor for profitability is also elaborated in a number of other studies. While the investigations in [49] still report moderate profits, battery wear is also mentioned as a relevant aspect affecting profitability. A study in [112] goes even further by concluding that peak shaving through V2G may result in a reduction of battery lifetime by approx. 3 years, thus leading to financial losses. Even more pessimistic are the conclusions drawn in [1] and a recent investigation in [2]. Using a battery aging model considering *charge rate* (C-rate), DOD and energy throughput, the authors in [1] arrive at the conclusion that using a PEV's battery for bulk energy services would result in the need for annual battery pack replacements. In [2] it is further argued that battery depreciation costs are highly likely to exceed expected revenues from V2G. Similar statements are made for the case of stationary batteries in [113] and [114] where the findings indicate that degradation costs considerably outweigh arbitrage revenues. Another study investigating the German market arrives at the conclusion that arbitrage profits could sum up to approx. € 250[5] [75]. This number, however, does not include battery degradation and it is

---

[5] In this work, a conversion rate of € 1 to US$ 1.11 is used.

argued that in a worst case scenario battery lifetime could be reduced to 3.7 years. A study published in [115] arrives at the conclusion that arbitrage in China is not economically feasible while the UK market may provide potential for annual profits in the range of approx. $ 40–275. Another comprehensive investigation on the example of the *Pennsylvania-New Jersey-Maryland Interconnection* (PJM) market can be found in [116]. In this work, the value of arbitrage using energy storage as a price-taker is estimated to $ 60–110 per kW and year for a round trip efficiency of 80 %. This study, however, does not capture battery degradation costs. Studies proposing smart charging approaches with the possibility of energy arbitrage show that only negligible amounts of energy are being discharged [117, 118]. An economic analysis using Singapore data can be found in [119]. In this study, it is concluded that energy arbitrage in Singapore is only economically feasible at particularly high real-time prices which were limited to about 100 time periods in 2012. This study, however, models battery degradation using fixed unit degradation costs per kWh which may only be achievable if the battery is cycled under very well-controlled conditions. The investigation is further based on average prices and therefore does not sufficiently account for the optimization potential.

This overview shows that studies concluding profitable operation mostly disregard battery degradation while investigations considering this aspect building on constant degradation costs usually predict financial losses. This clearly shows the relevance of the focus of this work which aims for profitable BESS operation by optimizing charge scheduling with regard to battery wear.

2.6.2.2 Ancillary Services

Participating in ancillary service markets is another possible revenue stream for BESSs. Apart from investigating bulk energy services, the study presented in [1] also analyzes ancillary services. For the latter, it is concluded that battery replacements every 2–3 years would be required. As a remedy, the authors suggest to reduce the battery capacity available for these services to avoid deep discharges. The economic analysis for the case of Singapore in [119] also investigates the economic viability of frequency regulation. Frequency regulation is found to be the most profitable service with annual profits of slightly more than S$ 1 800[6]. It is further concluded that profitability can be considerably increased by simultaneously participating in multiple markets. According to [120], income can be generated particularly in the primary reserve and the negative secondary reserve market with € 180 per year each in the German market. The same study, however, argues that the market volume for ancillary services is limited so that the German ancillary service market could be saturated by 2 million PEVs only. Another study for the German market presented in [3] concludes that average annual profits can range from € 360–960 with the highest profits in the secondary reserve market and the lowest ones for tertiary reserves. For the Swedish market, the same study states that no profits can be made. The considerable difference is due to the different capacity payments which at the time of writing were significantly lower in the Swedish market. In [23], a small-scale project conducted in the PJM market using PEVs as smart loads for the purpose of regulation services is presented where annual incomes in the range of $ 1 200–2 400 are achieved.

2.6.3 Battery Degradation Monetization

As discussed in Section 2.6.2, many studies identify battery degradation as an important cost driver. A weakness of most assessments, however, is that battery degradation costs are factored in as a constant. Assumptions are $ 0.3 per kWh in [121], € 0.03 per kWh for primary and secondary reserve and € 0.1 per kWh in the tertiary reserve market in [3], S$ 0.234 per kWh in [119] and $ 0.042 per kWh in [106]. This is a wide range with variation across two orders of magnitude which demonstrates the uncertainty related to quantify-

---

[6] In this work, a conversion rate of S$ 1 to US$ 0.72 is used.

ing this factor. Due to the important role of battery degradation costs for profitability, this wide spread has a high relevance for the validity of the conclusions.

Other approaches go one step further by accounting for battery degradation through limiting the SOC range [122–124] or restricting the C-rate [125–127]. Other recent approaches consider degradation as a function of the *SOC swing* Δx (sometimes alternatively denoted DOD) [128–132]. The static limitation of a battery's operation range without an underlying model, however, makes the restriction somewhat arbitrary and therefore does not lead to an optimal outcome. The one-dimensional dependency on the SOC swing better accounts for degradation while keeping the optimization problem simple. It nevertheless neglects the fact that degradation does not only depend on how deep a cycle is but also in what voltage range (i.e., around what SOC) it occurs. Furthermore, these simplifications neglect calendar degradation which in this work is also shown to be a relevant factor.

A practical battery wear model for PEV charging is proposed in [133]. It provides a generalized *wear density function* to derive cycle battery aging costs for any Δx based on data of the achievable cycle count for a specific DOD. The model, however, also neglects calendar aging processes. Another model is presented in [134] which serves for the prediction of cycle lifetime under a certain cycling regime. The model, however, also lacks the consideration of calendar aging and does not provide a framework for degradation monetization which would enable the integration into a scheduling model.

2.7 Conclusion

The considerations in this section outline the challenges and opportunities related to the integration of mobile and stationary BESSs into the power system. The steep decrease of battery prices which has repeatedly exceeded expectations may ultimately lead to a faster adoption of PEVs and stationary BESSs than currently predicted. This requires taking measures to facilitate the integration of these devices into the power system but also gives rise to new business cases employing BESSs for power grid services.

Despite decreasing prices, battery packs will remain a costly product and previous studies have shown that their deployment for power grid services can result in considerable depreciation costs. Operating batteries in a way which maintains their value in a best possible way is therefore one of the enabling factors for the economic competitiveness of PEVs and the deployment of BESSs in power systems. At the same time, however, the consideration of battery degradation for optimal scheduling and for assessing the economic viability of BESSs has not seen enough attention in the presented work. This work therefore puts a strong focus on considering battery degradation processes and on demonstrating the relevance of degradation monetization.

A notable body of research has investigated the economic feasibility of business cases employing stationary and mobile BESSs with widely varying results. This is because of the large number of parameters involved which vary both across different markets as well as over time. One important dimension in this context are electricity prices. Price levels, their intraday variances, forecasting horizons and forecasting accuracy play an important role for profitability considerations. Another important parameter are battery prices which have seen a steady decline in recent years. Conclusions therefore do not only have to be seen in the context of a particular market but also with regard to a certain observation period. In addition, the method applied to compute profits can have significant implications on the conclusions. While some approaches base their estimates on average prices, others employ TOU pricing or real-time pricing. The modeler also has a high degree of freedom in the choice of parameters such as battery capacities, connection power, energy losses and availability times which also have notable effects on the outcome.

One of the objectives of this work therefore is to provide a modular framework which can be employed to systematically investigate different scenarios by varying the types and properties of the ESS in use, by applying a certain setting to different markets and by providing the possibility for framework extensions in order to adapt to more advanced use cases. While it is clear that a single framework cannot replace the large variety of different control and scheduling approaches, it allows the quick investigation of the influence of a variety of design criteria. As this work is embedded in a context of the development of frameworks for the large-scale simulation of both traffic and power systems, it also allows considering mobility aspects relevant with regard to PEVs as well as investigating the power system impact of BESSs on a large scale. The combination of these frameworks therefore enables the investigation of a variety of topics from a holistic perspective.

3 Framework Architecture

3.1 Purpose

The objective of the presented framework is to provide a tool which can be employed to systematically investigate a variety of different scenarios involving ESSs including stationary BESSs and PEVs. A number of possible use cases of interest are given as follows:

- Investigation of the benefits of smart charging strategies with regard to electricity cost savings and battery wear for PEVs

- Assessment and comparison of the profitability of different types of ESSs in a particular electricity market and quantification of expected *return on investment* (ROI)

- Comparison of different electricity markets with regard to their profitability for a certain type of ESS

- Comparison of load zones or nodes within an electricity market with regard to their suitability for a particular type of ESS

- Identification of primary operating cost drivers of an ESS and their sensitivity with regard to the relevant parameters for different scenarios

- Identification of optimal ESS design parameters given certain market conditions

- Investigation on how certain operating conditions affect an ESS' lifetime

The framework may further be applied in combination with the other simulation frameworks CityMoS Traffic and CityMoS Power. This allows extending the range of topics to use cases such as:

© Springer Fachmedien Wiesbaden GmbH, part of Springer Nature 2019
D. Pelzer, *A Modular Framework for Optimizing Grid Integration of Mobile and Stationary Energy Storage in Smart Grids*, https://doi.org/10.1007/978-3-658-27024-7_3

- Quantification of operating costs of a PEV considering particular mobility patterns and charging behaviors for a population of drivers

- Investigation of the impact of certain PEV charging strategies or use of stationary ESSs on the power system

- Support of the simulation-based optimization of charging infrastructures for PEVs given certain price-responsive charging strategies

Since this work is motivated by issues related to PEVs, the models which are currently implemented in the framework focus on BESSs. As elaborated in Chapter 2, profit margins resulting from deploying BESSs for power grid services were found to be moderate, if existing at all. In two contributions [109, 110] prepared in the context of this work, it was therefore argued that two main aspects are crucial for profitable provision of power grid services. These are i) the exploitation of real-time prices and price forecasts for computing optimal charging and discharging schedules and ii) the explicit consideration of battery degradation as part of the optimization procedure. In the case of PEVs, a third aspect is equally essential which is the consideration of the driver's mobility pattern. The framework therefore provides the ability to solve multi-stage optimization problems given by time-series of real-time prices and price forecasts, it allows modeling and monetizing battery degradation behavior and in the case of PEVs to consider a driver's mobility pattern. Following the argumentation in Chapter 2 that a centralized control mechanism may be undesirable from a user perspective, both for privacy and financial reasons, the framework implements a decentralized approach in which decisions are made by the individual BESS. The implemented concept is essentially a *model predictive control* (MPC) approach [135]. This means that the behavior of the system is optimized based on information and models predicting the development of the system state with a certain *lookahead*. As information may be updated and new information may become available as time progresses, a decision is always made only for the current time period based on the computed *look-ahead policy*. At each time step, a new lookahead policy is computed and the decision for the first time step is implemented. The implementation of this approach is described in detail in Section 4.2.

To be able to address the above-mentioned use cases in a most generic manner, the primary design criteria of the framework are modularity and extensibility. As described in Section 3.2, the framework is therefore divided into functionally different modules; code and input data are strictly separated. This allows changing parametrizations of implemented models and extending or replacing individual models in accordance with the requirements for answering a particular question. While the current implementation focuses on Li-ion batteries for the reasons outlined in Chapter 2, the battery pack module can therefore simply be replaced by a module describing a different ESS technology such as a supercapacitor, a flywheel or a thermal storage device without affecting the functionality of other framework components. Similarly, individual parts of the framework such as the solver or a battery model can be used for purposes outside the current implementation.

In Section 3.2, a high-level overview of the framework is provided. Sections 3.3 and 3.4 provide the details of the framework components and their interactions. In Section 3.5, the other simulation frameworks CityMoS Traffic and CityMoS Power also used in this work are presented. Section 3.6 concludes this chapter.

3.2 Framework Overview

The framework strictly separates source code, configuration parameters, input data and output data to allow simple adaptation to different markets, manipulation of parameters and automated parameter sweeps. This section provides a brief overview of the various parts of the framework, the different components are then described in greater detail in Section 3.3. At a very high level, the following parts can be distinguished:

- **Source code**
  The source code is structured into various modules which implement different functionalities. This allows either deploying the framework as a whole or using single modules individually for different applications. Since the framework is implemented in Python, each module corresponds to a Python file defining its own namespace. A module may contain one or multiple classes. To formally distinguish between modules and classes, this presentation uses lower case names with underscores for modules and PascalCase for classes. The individual modules are discussed in more detail in Section 3.3.

- **Configuration**
  All the configuration data is described by using XML which encodes documents in a structured way easily readable for both humans and machines [136]. To extend the concept of modularity to the configuration, distinct XML documents are associated with each module. A module may be accompanied by a single or a hierarchical set of XML documents as further elaborated in Section 3.3. The interface between source code and configuration is provided through XML parsers contained in a `data_handler` module also described in Section 3.3. An XML document is *well-formed* if its syntax follows the XML specification and *valid* if it is well-formed and at the same time complies with an XML Schema. The latter can be specified by an *XML Schema Definition* (XSD) [137, 138] which formally defines a particular XML document's syntax. In the given context, XSD files can allow the developer of a module to ensure that only valid configuration parameters are accepted and to enable the user to add modified data in accordance with these specifications.

- **Input data**
  In the given context, the input data consists of electricity mar-

ket data. All market data is stored in a PostgreSQL database which is structured in accordance with the electricity market characteristics introduced in Section 2.3. The corresponding database schema is provided in Figure 3.1. For each **Market** specified by its **name**, **country** and the market **operator**, there may be several submarkets. A **Submarket** addresses a particular **market_type** such as a fixed price, day-ahead or real-time market. Furthermore, the **service_class** such as retail, generation, regulation, primary, secondary and contingency reserve is specified. Each submarket also has a certain maximum spatial resolution **max_resolution_spatial**, such as *zonal* or *nodal* as well as a particular time period during which prices are constant **max_resolution_temporal**. For the latter, the resolution is given in units of minutes. The table **Location** contains the **name** of each location the market provides price data for as well as the **type** of this location which may be a node, zone or the entire market area. For each submarket, there may be multiple price data sets which are specified in the table **Price_data_set**. The price contained in a data set has a name **name_price** specified by the official market description and a spatial and temporal resolution **resolution_spatial** and **resolution_temporal** which cannot be greater than the maximum resolution specified by its related submarket object in the **Submarket** table. The **payment_type** specifies whether the remuneration is based on a payment for energy or capacity. The **service_direction** indicates whether the price concerns the delivery or purchase of energy if the **payment_type** is specified as energy, or the provision of up or down regulation capacity if the **payment_type** is specified as capacity. The forecasting horizon for the particular price is given by the **forecast_lookahead** with a value of 0 indicating a cleared price. A value of 1 describes a price which is forecasted for one period in the future and so on. As the names say, **currency** and **datasource** indicate the currency of the data set as well the source from which it was obtained. Finally, a **Price** consists of a **datetime** attribute, the time period indicators

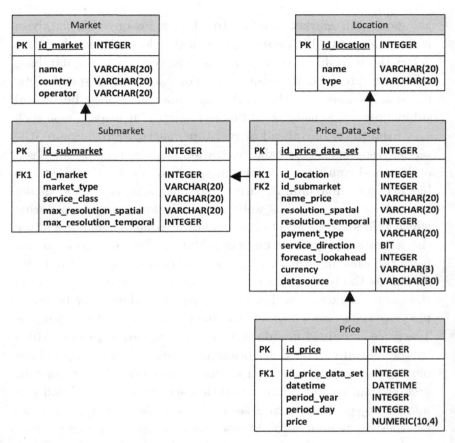

Figure 3.1: Price database scheme containing the required market information.

period_year and period_day and the actual value of the price in the
specified currency.

- **Output data**
  Similarly to market input data, output data is contained in a PostgreSQL database for easy structuring and analysis. The output
  database is simple with only two tables. The first one stores all
  simulation parameters corresponding to a particular simulation
  run with the id of the experiment id_run as primary key. In the

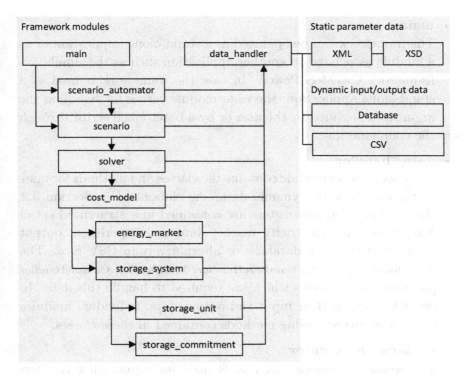

Figure 3.2: Framework architecture overview.

second table with `id_run` as a foreign key, the results computed for each individual time period are stored. For easier debugging, output data can also be written in CSV format.

3.3 Components

The framework consists of various components for controlling the program flow, handling data, conducting the computations related to the individual models and solving the optimization problem. An overview of the framework is shown in Figure 3.2 and the functions of the various packages are outlined as follows:

- **main**

  The framework can be operated as a stand-alone application or as a module associated to another application such as the simulation framework CityMoS Power. In case the framework is used as a stand-alone application, the main module serves for handling the arguments provided by the user or by a batch/shell script through the command line.

- **data_handler**

  The data_handler module contains the classes and methods for managing any static or dynamic data. As elaborated in Section 3.2, the configuration parameters are contained in a structured set of XML files while electricity market data and experiment output data are stored in a database or alternatively in CSV files. The data_handler provides **DataBaseHandler**, **XmlHandler**, **ConsoleHandler** and **CsvHandler** classes which are required to handle this data. In order to manage their input and output data, individual modules access the corresponding methods contained in these classes.

- **scenario_automator**

  The scenario_automator module is used for automating the execution of different scenarios. This allows conducting parameter sweeps and deploying different combinations of models by specifying ranges or lists of values. For this purpose, the methods in the **ScenarioAutomator** class contained in this module parse arguments received from main to automatically create the required scenarios.

- **scenario**

  The scenario module specifies the scenario to be investigated. If the framework is used as a stand-alone application, the instructions for setting up a scenario are obtained from the main module. In case the framework is used as part of a different application, this application may directly import the scenario module to create a **Scenario** instance. The **Scenario** class manages the execution of the optimization process and ensures that all state variables are updated as time proceeds and the system state changes. The basic

Scenario class can be extended in order to implement new types of scenarios. In the case of the MultiStageScenario which is of primary interest in this work, it divides the problem into sub-problems which can be processed by the Solver contained in the solver module. For this purpose, it extracts the relevant state variables from the deployed StorageSystem instance as provided by the storage_system module. It furthermore collects information on possible commitments of the storage system from a StorageCommitment object in the storage_commitment module. This information is turned into state and control variable constraints which are part of the arguments for initializing the Solver. After solving the optimization problem, the instance of the Solver class hands the solution of the sub-problem back to the Scenario instance which updates the state of the StorageSystem instance accordingly. The relevant output data is then written to the database using the appropriate methods provided by the data_handler module. Finally, the next sub-problem to be solved is set up. In an MPC setting, typically only the solution for the first time step is used because new information becomes available as time proceeds. A new sub-problem is therefore solved for the subsequent time period using updated information. An exemplified definition of the configuration file for a Scenario is provided in Listing 3.1.

- **solver**
  The task of the solver is to minimize a utility function given a particular system state as well as certain constraints. The setting considered in this work is a multi-stage optimization problem consisting of a series of *stages* and a discrete number of *states* which are accessible at a particular stage. In the given context, stages represent time steps while the states correspond to the SOC of the ESS. The utility function is composed of the revenues which can be attained in various kinds of electricity markets as well as the cost aspects related to ESS operation and energy purchase. The mathematical details concerning the optimization problem and the

Listing 3.1 Exemplified Scenario specifications.

```
 1  <Scenarios>
 2    <Scenario id="1">
 3      <StorageSystem>
 4        <!-- STORAGE SYSTEM TO USE -->
 5        <!-- REFERS TO SPECIFICATIONS IN A StorageSystem XML
             FILE-->
 6        <!-- STORAGE SYSTEM INITIAL STATE -->
 7      </StorageSystem>
 8      <Market>
 9        <!-- MARKET/SUBMARKETS/PRICE TYPES TO USE -->
10        <!-- REFERS TO SPECIFICATIONS IN A Market XML FILE-->
11      </Market>
12      <Commitment>
13        <!-- COMMITMENTS, E.G., MOBILITY PATTERN TO USE -->
14        <!-- REFERS TO SPECIFICATIONS IN A CommitmentModel
             XML FILE-->
15      </Commitment>
16      <Solver>
17        <!-- SOLVER TO USE -->
18        <!-- REFERS TO SPECIFICATIONS IN A Solver XML FILE-->
19      </Solver>
20      <Experiment>
21        <!-- EXPERIMENT SETTINGS, E.G., TIME HORIZON -->
22      </Experiment>
23    </Scenario>
24    <!-- ... -->
25  </Scenarios>
```

solution algorithm are described in detail in Chapter 4.

An instance of the Solver class receives the current system state and a set of constraints from a Scenario instance. With regard to the constraints, it can be distinguished between *control constraints* representing conditions for the value of the control variable and *state constraints* which may limit the range of accessible states. Control constraints may, for instance, result from ESS specifications limiting charging or discharging rates or from the existence of an external load which is drawing power from the device. If the ESS is a component of a PEV, an example for state constraints is a minimum SOC threshold which may be required by the driver as a safety buffer for unexpected energy use. Using these inputs, the Solver instance minimizes the aggregated cost over a defined number of stages defined by the lookahead to find an optimal charge/discharge policy. For this optimization, the Solver instance requires the cost associated to each possible state transition at each stage. This cost is computed in an instance of the CostModel class in the cost_model module.

The solver is implemented as a *dynamic programming* (DP) algorithm which is particularly suitable for addressing multi-stage optimization problems where system states can be discretized. All required details on the DP approach are presented in Section 4.2.1. Since the DP algorithm operates in a discrete space, certain calculations are repeated a great number of times in an exactly similar manner which allows building up lookup tables on the go for speeding up computations. Furthermore, computations within each stage are independent from each other which allows distributing the load over multiple processes to make use of multi-core or multi-node architectures. As the solver module operates solely on the basis of generic state and control variables as well as corresponding constraints, it is implemented in a generic way so that it is also applicable for entirely different sorts of problems. The definition of the simple Solver specifications is sketched in Listing 3.2.

- **cost_model**
  An instance of the Solver class requires a cost for each state transition at every stage. The calculation of this cost from all relevant cost drivers is performed by the cost_model module which implements a CostModel class. The Solver instance therefore passes a vector describing a particular control decision to the CostModel instance, expecting to receive the cost related to this control decision. In the particular setting under consideration, costs and revenues incur from buying or selling energy or from providing power capacity to the PSO. To quantify these costs, the CostModel instance requires energy prices as provided by the energy_market module. Additional cost factors are given by energy losses as well as by

Listing 3.2 Exemplified **Solver** specifications.

```
 1  <Solvers>
 2    <Solver name="dynamic programming">
 3      <setting id="1">
 4        <!-- STATE SPACE SETTINGS -->
 5        <xMin>0</xMin>
 6        <xMax>1</xMax>
 7        <dx>0.05</dx>
 8        <!-- PERFORMANCE SETTINGS -->
 9        <useMultiThreading>y</useMultiThreading>
10        <useLookupTables>y</useLookupTables>
11        <resetLookupTablesInterval>200</
              resetLookupTablesInterval>
12      </setting>
13      <setting id="2">
14        <!-- ... -->
15      </setting>
16    </Solver>
17    <!-- ... -->
18  </Solvers>
```

operating costs of the ESS. To allow for integrating any kind of
ESS including batteries, supercapacitors, hydro-storage, flywheels,
thermal storage etc., the computation of the physical quantities
which determine the cost of a certain decision is performed in the
storage_system module. The **CostModel** then monetizes these physical dimensions such as a change in the remaining lifetime of the
ESS. For adapting the way of monetizing these physical dimensions, the respective methods provided by the **cost_model** module
have to be overridden. The **CostModel** in its current implementation does not require any parameters which is why the corresponding XML document is empty.

- **energy_market**
  In an indirect control setting, the controller of an ESS may interact
  with various electricity markets such as retail or wholesale markets
  through prices as described in Section 2.3. In many practical cases,
  the entity providing the prices may also be an intermediary such
  as a VPP operator. A **CostModel** instance requires price data to

compute the cost of a control decision. This information is provided through the energy_market module. In the presented setting, cleared prices are used and the ESS operator is considered a price taker. The role of the energy_market module in this case therefore is simply to provide the appropriate real-time prices and price forecasts for the particular time of interest. For a real-time application, the energy_market module could also implement an interface querying price data from the market operator. It may further be extended by including price forecasting mechanisms or may represent an exchange market in case the ESS is not treated as a price taker. The configuration for instantiating the Market class in the energy_market module is provided by an XML document which is exemplified in Listing 3.3. The specifications follow from the description of electricity markets introduced in Section 2.3.

- **storage_system and storage_unit**
  In order to compute costs for the Solver instance, the CostModel instance requires information on the physical state of the ESS. Since the Solver operates in a discrete state space in which the states correspond to the SOC of the ESS, the cost_model also requires information on the power and the amount of energy corresponding to a certain state transition. An ESS exhibits a variety of state variables such as energy or charge content, power rating, efficiency etc. These are modeled in the storage_system module. For certain technologies such as a battery pack, an ESS may consist of a number of storage units, e.g., battery cells. These may be described by additional sets of variables such as voltage and physical health parameters, all of which may be subject to change on various time scales. This requires additional sophisticated models to describe the behavior of a storage unit. The attributes and possible actions of a storage unit are modeled in the module storage_unit.
  The concept of the storage_system and storage_unit is illustrated in Figure 3.3. The general function of an ESS, irrespective of the technology it implements, is the ability to absorb and deliver a

Listing 3.3 Exemplified **Market** specifications.

```
 1  <Markets>
 2    <Market name="nems" country="singapore" operator="emc">
 3      <Submarket id="1">
 4        <marketType>rt</marketType>
 5        <serviceClass>generation</serviceClass>
 6        <maxResolutionSpatial>nodal</maxResolutionSpatial>
 7        <maxResolutionTemporal>30</maxResolutionTemporal>
 8        <!-- ... -->
 9      </Submarket>
10      <Submarket id="2">
11        <!-- REGULATION MARKET ATTRIBUTES -->
12      </Submarket>
13      <Submarket id="3">
14        <!-- PRIMARY RESERVE MARKET ATTRIBUTES -->
15      </Submarket>
16      <Submarket id="4">
17        <!-- SECONDARY RESERVE MARKET ATTRIBUTES -->
18      </Submarket>
19      <Submarket id="5">
20        <!-- CONTINGENCY RESERVE MARKET ATTRIBUTES -->
21      </Submarket>
22      <!-- ... -->
23    </Market>
24    <Market name="pjm" country="usa" operator="pjm
          interconnection">
25      <!-- PJM MARKET ATTRIBUTES -->
26    </Market>
27    <!-- OTHER MARKETS -->
28  </Markets>
```

certain maximum amount of energy at a certain rate and with
a particular efficiency. In the **storage_system** module, this storage
system is implemented by the class **StorageSystem**. In the presented
case, the **StorageSystem** is implemented as a **BatteryPack** which extends the **StorageSystem** class. An instance of the **BatteryPack** class
is further composed of battery cell instances implemented by the
class **BatteryCell** which extends the class **StorageUnit** contained in
the **storage_unit** module. The physical properties of the **BatteryCell**
including voltage response and degradation behavior are specified by models which are described in Sections 4.3 to 4.5. Examples for XML documents which describe the parameters of the

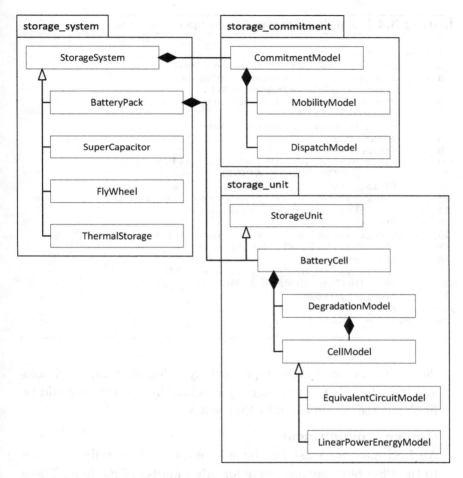

Figure 3.3: Overview of the classes of the modules **storage_system**,
storage_unit and **storage_commitment**.

StorageSystem, **StorageUnit**, a **CellModel** and a **DegradationModel** are
provided in the Listings 3.4 to 3.7.

The implemented models are applicable to a certain type of Li-ion
battery, however, by using different parametrizations or replacing individual sub-models, different battery types may be implemented. Since the other components of the framework are indepen-

Listing 3.4 Exemplified StorageSystem specifications.

```
1   <StorageSystems>
2     <StorageSystem id="1" type="batteryPack">
3       <idStorageUnit>1</idStorageUnit>
4       <capacityKWh>20</capacityKWh>
5       <investmentCosts>150</investmentCosts>
6       <Inverter>
7         <efficiency>0.98</efficiency>
8       </Inverter>
9     </StorageSystem>
10    <StorageSystem id="2" type="batteryPack">
11      <!-- BATTERY PACK 2 ATTRIBUTES -->
12    </StorageSystem>
13    <StorageSystem id="3" type="superCapacitor">
14      <!-- SUPER CAPACITOR ATTRIBUTES -->
15    </StorageSystem>
16    <StorageSystem id="4" type="flyWheel">
17      <!-- FLY WHEEL ATTRIBUTES -->
18    </StorageSystem>
19    <StorageSystem id="5" type="thermal">
20      <!-- THERMAL STORAGE ATTRIBUTES -->
21    </StorageSystem>
22    <!-- ... -->
23  </StorageSystems>
```

dent of the actual storage type, entirely different storage technologies including thermal storage, flywheels or hydro-storage could be implemented as indicated in Figure 3.3.

- **storage_commitment**
  An ESS may not exist for the sole purpose of providing services to the PSO but may also serve for other kinds of functions. These can include buffering power generation of a residential PV installation, providing power to a residential load or serving for driving purposes of a PEV. These applications may manifest themselves as constraints to the Solver or require a co-optimization as additional terms of the utility function. Equivalently to electricity prices, the power consumption or generation resulting from these loads and generators can typically be forecasted to a certain degree, e.g., using external forecasts or internal time series models. The pur-

Listing 3.5 Exemplified `StorageUnit` specifications.

```
 1  <StorageUnits>
 2    <StorageUnit id="1">
 3      <idCellModel>1</idCellModel>
 4      <idDegradationModel>1</idDegradationModel>
 5      <storageUnitType>batteryCell</storageUnitType>
 6      <cellChemistry>nmc</cellChemistry>
 7      <manufacturer>panasonic</manufacturer>
 8      <ratedCapacity>2.15</ratedCapacity>
 9      <nominalVoltage>3.6</nominalVoltage>
10      <minVoltage>2.7</minVoltage>
11      <maxVoltage>4.2</maxVoltage>
12      <maxChargeRate>1</maxChargeRate>
13      <maxDischargeRate>-2</maxDischargeRate>
14    </StorageUnit>
15    <StorageUnit id="2">
16      <!-- BATTERY CELL 2 ATTRIBUTES -->
17    </StorageUnit>
18    <!-- ... -->
19  </StorageUnits>
```

Listing 3.6 Exemplified `CellModel` specifications.

```
 1  <CellModels>
 2    <CellModel id="1">
 3      <ocvParameters>
 4        <!-- OPEN CIRCUIT VOLTAGE PARAMETRIZATION -->
 5      </ocvParameters>
 6      <ohmicResistanceParameters>
 7        <!-- OHMIC RESISTANCE PARAMETRIZATION -->
 8      </ohmicResistanceParameters>
 9      <!-- ... -->
10    </CellModel>
11    <!-- ... -->
12  </CellModels>
```

Listing 3.7 Exemplified **DegradationModel** specifications.

```
1   <DegradationModels>
2     <DegradationModel id="1">
3       <calAgingParameters>
4         <!-- CALENDAR DEGRADATION EQUATION PARAMETRIZATION
            -->
5       </calAgingParameters>
6       <cycAgingParameters>
7         <!-- CYCLE DEGRADATION EQUATION PARAMETRIZATION -->
8       </cycAgingParameters>
9     </DegradationModel>
10    <!-- ... -->
11  </DegradationModels>
```

pose of this module is to provide an environment to model the commitments the ESS has with regard to these additional loads or generators. In particular, this requires a forecast of the power the ESS has to provide or consume during a certain forecasting horizon. This information is provided by the storage_commitment module.

In the presented application, a relevant use case is given by PEVs. While a vehicle is used for driving, it is not available for power grid services and energy is drawn from the battery. As indicated in Figure 3.3, the PEV's availability is determined by the MobilityModel class extending the CommitmentModel class which is part of the storage_commitment module. In the presented case, the mobility behavior is given by deterministic mobility patterns which are provided by the traffic simulation CityMoS Traffic. The MobilityModel in its current implementation therefore only has to manage this data. This includes the discretization of mobility data in both time and space as it is required by the solver. An example XML document for specifying mobility pattern for the MobilityModel is shown in Listing 3.8.

Listing 3.8 Exemplified MobilityModel specifications.

```
 1  <CommitmentModel>
 2    <MobilityModel>
 3      <pattern id="1">
 4        <!-- MOBILITY PATTERN ATTRIBUTES (TIME INTERVAL/POWER
             CONSUMPTION) -->
 5      </pattern>
 6      <pattern id="2">
 7        <!-- MOBILITY PATTERN ATTRIBUTES (TIME INTERVAL/POWER
             CONSUMPTION) -->
 8      </pattern>
 9      <!-- ... -->
10    </MobilityModel>
11    <ResidentialLoadModel>
12      <!-- RESIDENTIAL LOAD PATTERNS WOULD BE SPECIFIED HERE
           -->
13    </ResidentialLoadModel>
14  </CommitmentModel>
```

3.4 Interactions

To clarify its operating mechanisms, a high-level interaction sequence of the framework when used as a stand-alone application is shown in Figure 3.4. According to this presentation, the typical interaction sequence looks as follows:

1. The program flow starts at Main which processes the received command line arguments and creates a new optimization Scenario.

2. A new scenario requires instances of the various classes Solver, CostModel, Market and StorageSystem.

3. In the Scenario, a sub-problem consisting of the stages i to $i + l$ with l denoting the lookahead is split off the overall optimization problem and passed to the Solver.

4. The Solver proceeds through all stages of the sub-problem from $k = 0$ to l computing the costs for the transitions between all states. This requires the CostModel. The CostModel needs the

Market as well as the StorageSystem to compute all relevant cost factors including energy costs and battery degradation costs.

5. After computing the optimal path, the Solver hands the solution to the sub-problem back to the Scenario. This uses the solution for the first stage, updates the storage system accordingly, writes the result to the database using the appropriate method and proceeds again with Step 3.

6. After the entire optimization horizon has been covered, the program is terminated in Main.

For greater clarity, the data_handler is not shown and the possibility for automating scenarios is omitted in Figure 3.4.

3.5 Transportation and Power System Simulation

The framework was developed and deployed in the context of the frameworks CityMoS Traffic and CityMoS Power. Both of these frameworks were used for the investigation of the power grid impact of PEV charging presented in Section 2.5.2.3. The studies presented in Chapter 5 are partly built on simulations conducted with CityMoS Traffic. Therefore, this section provides a brief overview of these frameworks. For further details, the reader is referred to the explanations in [6–8, 10, 139].

3.5.1 CityMoS Traffic

CityMoS Traffic is an agent-based *discrete event simulation* (DES) framework designed to run on *high performance computing* (HPC) infrastructures. Its purpose is to simulate traffic at the scale of a megacity down to the level of individual agents. This shall enable the simulation-based exploration of a large variety of different scenarios resulting from aspects like infrastructure modifications, implementation of novel traffic control measures and integration of autonomous vehicles into the transportation system etc. In the broader

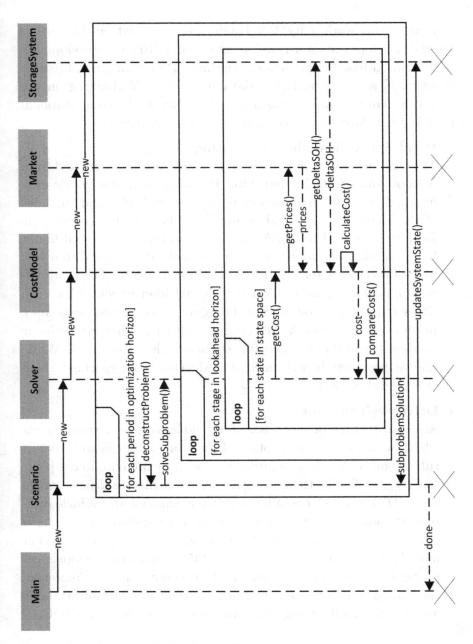

Figure 3.4: Conceptualized sequence diagram.

context of this work, CityMoS Traffic was employed by this author for the development of a ridesharing concept [10], the development of simulation-based approaches to optimize PEV charging infrastructures [8, 9] and to investigate the impact of PEV charging on the power system [5]. As described in the following, the computational model of CityMoS Traffic consists of three main parts:

- **Road network model and routing**

  The road network is implemented as a directed graph created from data obtained from Navteq [140]. It is composed of edges describing road *lanes* and nodes specified by the longitudes and latitudes of the lanes' start and end points. The data further contains information on road properties such as road types and speed limits. Lanes sharing the same start and end node are assigned to a *link* which describes a road with one or multiple lanes. The road network model contains traffic lights at 200 intersections which are either controlled through fixed timings or based on link occupancy. Routes on this network are computed either as shortest paths or shortest travel times employing *Dijkstra's* algorithm [141]. While vehicle movement is performed at the level of lanes, routing is done at the level of links.

- **Driver-vehicle unit**

  An agent-based model consists of individual entities termed *agents* which have a certain perception of their surroundings and a set of rules to make decisions in interaction with their environment [142]. In CityMoS Traffic, these agents are implemented as *driver-vehicle units* (DVUs) which contain both driver behavior and vehicle component models. Driver models include car-following and lane-changing, acceleration and deceleration, overtaking, intersection and charging behavior models [143–145]. Component models describe the drive train, battery and air-conditioning. Combined, these models determine a vehicle's movement and inform about aspects such as its energy consumption and the battery's SOC.

- **Traffic assignment**

  An agent's trip is characterized by a pair of coordinates describing its origin and destination as well as a starting time. In the Singapore context where the simulation is used, trips can start and end at any of the 6-digit postal codes. The origin-destination matrix for generating all trips of the day is created based on data of the *Household Interview Travel Survey* (HITS) [146] of the year 2012 which captures the mobility behavior of 10 000 Singapore households. As this data set only represents a fraction of the population, a temporal and a spatial extrapolation are performed according to the procedure described in [8] in order to generate the desired number of agents. Agents can have multiple trips per day; a combination of trips is termed an *itinerary*.

3.5.2 CityMoS Power

As described in further detail in [5–7], CityMoS Power is a power system simulation framework designed for synthesizing and evaluating PNMs as well as for conducting power flow simulations. This work contributed to the development of this framework as presented in [7]. Both components of the framework were used to investigate the power system impact of PEV charging in a Singapore context as presented in Section 2.5.2.3. This section therefore provides a brief overview of both the PSP and the power flow simulation component. For an exhaustive presentation, the reader is referred to [6].

3.5.2.1 Power System Planning

Investigating the effect of ESSs on power system infrastructures requires the ability to conduct tempo-spatially resolved simulations of the power flow on a system scale. One crucial input for these simulations is detailed information on the power network including the locations and capacities of generators, consumers, substations and power lines. Publicly available data on these aspects is typically sparse which makes conducting this sort of studies difficult. This issue is addressed by the PSP module in CityMoS Power which allows synthesizing realistic PNMs based on publicly available data.

The power network synthesis is based on tempo-spatially resolved data of the electrical load as well as on a number of assumptions on power grid topologies. The synthesis procedure starts at the lowest voltage level by creating geographic areas with similar power demand using a weighted *k-means* algorithm [147]. A substation is then placed at the load center of each area. Within this area, demand points shall be connected to the substation. Generators are generally referred to as *PV buses* and loads as *PQ buses*. A substation is considered a PV bus if seen from a lower voltage level to which it provides power. From the perspective of the higher voltage level, a substation is considered a PQ bus since it acts as a consumer from this point of view. In order to connect demand points to substations, the *density-based spatial clustering of applications with noise* (DBSCAN) algorithm [148] is used to perform a density-based clustering of these buses. Depending on the desired network topology, buses in each cluster are connected by *branches* in different ways. For rings, as they are common in the LV and MV grid, this is done by solving a *traveling salesman problem* (TSP) [149] starting and ending at an area's substation. Equivalently, radial networks as they are most common in the MV layer are generated by creating a *minimum spanning tree* (MST) [150]. Meshed networks as most common in the HV grid are generated by employing a *Delaunay triangulation* (DT) [151].

For each area to be technically feasible, three main conditions need to be fulfilled:

1. The aggregated power demand of all N consumers and the power losses in all M branches within this area must not exceed the maximum power $P_{S,\max}$ the corresponding substation can supply.

2. The total power flow P_j through each branch j must not exceed the nominal power rating $P_{\max,j}$ of the respective branch so that $P_{\max,j} \geq P_j$ for each j.

3. The voltage magnitude $|V_i|$ at each consumer i must be within a predefined range given by $|V_{\min}| \leq |V_i| \leq |V_{\max}|$.

If one of these conditions is violated, the corresponding area is split up into two, the first substation is placed in the center of the first area and an additional substation is placed in the center of the second one. After all areas in the lowest voltage layer have been planned, the algorithm continues at the next higher voltage level. For synthesizing this layer, the previously created substations complement the set of existing consumer buses. These points are then again grouped into clusters and connections between consumer buses and supply stations are established. Repeating this procedure for each voltage layer up to the highest level finally leads to the desired PNM.

3.5.2.2 Power Flow Simulation

The power flow simulation computes the capacity utilization at every branch as well as the voltage and phase angle at every bus. Calculations may cover an arbitrary period of time and are discretized according to predefined time intervals. The result informs about the current state of the power grid and identifies times and locations of grid congestion and voltage drops.

The power flow model is based on the AC power flow simulation *JPOWER* [152] which uses the *Newton-Raphson* algorithm for analyzing the load flow [153]. To reduce the computational requirements, calculations are performed consecutively for independent parts of the grid such as distinct rings or branches. The outputs generated for one voltage layer are then taken as inputs for the next level.

The simulation is initialized by defining a substation corresponding to a ring or radial branch as slack bus and by declaring each consuming bus i a PQ bus. PQ buses are initialized by their respective active and reactive power demand, $P_{D,i}$ and $Q_{D,i}$, while voltage magnitudes $|V_i|$ and phase angles Θ_i are initialized with $1.0\angle 0°$ pu.

The power flow simulation starts at the lowest voltage level by determining the active and reactive power on both ends of every branch within a ring. The power difference between both ends of a branch j is thus the power loss $P_{L,j}$ and $Q_{L,j}$. Total active and reactive power in one ring are then calculated according to

$$P_{S} = \sum_{i=1}^{N} P_{D,i} + \sum_{j=1}^{M} P_{L,j} \tag{3.1}$$

$$Q_{S} = \sum_{i=1}^{N} Q_{D,i} + \sum_{j=1}^{M} Q_{L,j} \tag{3.2}$$

After calculating the power flows within one voltage layer, the process is repeated for the next higher level. For this purpose, $P_{D,i}$ and $Q_{D,i}$ at each substation in the higher level are set to the previously calculated values $P_{S,i}$ and $Q_{S,i}$ of the corresponding substation in the lower voltage winding. It is assumed that shunt capacitor banks are installed at each station. This justifies setting the initial values for $|V_i|$ and Θ_i to $1.0\angle0°$ pu for each voltage level.

If at any PQ bus the voltage magnitude violates the third condition defined in Section 3.5.2.1, another simulation iteration is initiated by increasing $|V_i|$ at the slack bus up to $|V_{\max}|$. In case this condition still cannot be satisfied, the corresponding part of the grid is considered overloaded.

3.6 Conclusion

This chapter presents the computational frameworks developed and deployed in the context of this work. The general paradigm underlying the scheduling framework is universal applicability and extensibility. For this purpose, a modular approach is pursued which allows replacing, extending or adding individual models. This allows the systematic and automatic investigation of different scenarios related to the grid integration of ESSs with a variety of parameter and model combinations. The same paradigm underlies the development of the simulation platforms CityMoS Traffic and CityMoS Power which are designed building on a systemic approach attempting to understand the behavior of the system based on the interactions between the system's components. This allows modeling emergent phenomena which are difficult to address from a top-down perspective.

While the universal approach allows addressing a great variety of use cases, it also has to be acknowledged that in accordance with the *no free lunch theorems* [154], applicability to a wide range of problems generally comes at the cost of performance for individual problems. The presented approaches are therefore very well suited for exploring large scenario spaces in order to narrow down system design parameters, to establish an understanding of a system's behavior and to compare different scenarios with each other. For accurate solutions or investigations addressing hard optimization problems requiring high performance, however, a universal approach is likely to reach its limits. In these cases, the presented approaches can be employed to gain a high-level understanding of the system in order to develop methodologies tailored to the particular problem.

The limitations of the scheduling framework are primarily a consequence of the limitations of the implemented models which are discussed in Chapter 4 and in the context of the possibilities for future work in Chapter 6.

With regard to CityMoS Traffic, a limitation at the time of writing was the restricted number of traffic lights and the lack of traffic light control schemes simulating the *Sydney Coordinated Adaptive Traffic System* (SCATS) [155] which is implemented in Singapore. As traffic light control has a major impact on traffic flows, this can be expected to result in a notable difference between real and simulated traffic. These limitations further result in the need to restrict the number of vehicles on the roads below a realistic number in order to avoid grid locks. The restrictions have direct implications on trip durations and, as most relevant in the context of this work, energy consumption. The results building on computations for energy consumption therefore need to be seen in this context as further discussed in Chapter 5. Due to the large number of traffic lights in urban transportation systems and as a consequence of the fact that traffic control is still often manually devised by optimizing the flow across individual sets of critical interjunctions, this issue remains difficult to address on a large scale.

The limitations regarding CityMoS Power were discussed in the context of the power grid impact study in Chapter 2. These include the greenfield strategy generating a PNM from scratch, thus not accounting for the evolution of power systems which typically emerge through incremental expansion. Accounting for this aspect is prohibitively difficult since it would require knowledge on the historical tempo-spatial power demand including historical demand projections which served system developers for their planning decisions. The mentioned lack of considering geographical constraints when placing substations and power lines would be easier to address by feeding the PSP process with information obtained through a *geographical information system* (GIS). Finally, the assumptions of homogeneous specifications for substations and power lines could be relaxed if more real-world data was available. As system operators are generally restrictive with regard to information on power infrastructures, even this information, however, is difficult to obtain.

A limitation of the power flow simulation is that it relies on network decomposition, meaning that the power flows in different areas are simulated individually. This is required for performance reasons to be able to conduct the AC power flow simulations in a reasonable amount of time. With regard to real-world networks, this is a simplification since these contain redundancies connecting distinct parts of the network with each other. For the predominant radial and ring network topologies, however, these redundancies play a minor role which is why this limitation is considered acceptable for the presented use cases.

4 Scheduling Approach

4.1 Introduction

In Section 2.7, it was concluded that profitable operation of a BESS depends on real-time prices, electricity price forecasts, battery aging characteristics and possibly further constraints as they may result from a PEV owner's driving patterns or other sorts of connected loads. This chapter presents the implemented approach for BESS scheduling accounting for these aspects. The problem under consideration is a multi-stage optimization problem where an optimal charging/discharging schedule has to be computed given a time-series of electricity prices and constraints. This scheduling approach is introduced in Section 4.2. As the control and state variables in this section are not necessarily tied to a BESS, the presented optimization concept is also applicable to different kinds of ESSs. The connection between the control and state variables defined in Section 4.2 for the particular case of a BESS is therefore established in the following Sections 4.3 and 4.4. From the perspective of the scheduling mechanism, the quantity of interest is the power at the terminal of the BESS while at cell level charging or discharging processes are characterized by the movement of charges between the electrodes. Relating these aspects with each other requires a *battery model* which is presented in Section 4.3. Based on this model, a charging model can be established which connects the SOC change with the power measured at the terminal during charging or discharging. This model is presented in Section 4.4. As the cell ages, a number of parameters undergo changes which manifest themselves on a macroscopic level, such as through capacity fade and internal resistance increase. These have an effect on the performance of the BESS and the change of these quantities is closely tied to operating costs. The aging trajectory of the cell notably depends on the operating conditions which in turn are a result of the scheduling strategy. To optimize the scheduling

© Springer Fachmedien Wiesbaden GmbH, part of Springer Nature 2019
D. Pelzer, *A Modular Framework for Optimizing Grid Integration of Mobile and Stationary Energy Storage in Smart Grids*, https://doi.org/10.1007/978-3-658-27024-7_4

with regard to battery wear, this requires a *battery degradation model*
which quantifies and monetizes the cell's aging behavior as a function
of the scheduling method's control variables. A model fulfilling this
purpose is presented in Section 4.5. A conclusion of this chapter is
provided in Section 4.6.

4.2 Multi-Stage Optimization Problem with Rolling Horizon

In general terms, a scheduling problem is the problem of minimizing
a certain objective function by carrying out a particular number of
tasks at specific times. Consequently, the solution of the scheduling
problem defines a schedule describing a series of tasks which have to
be executed in order to achieve an optimal outcome. In the given case,
the goal is to find an optimal sequence of charging/discharging actions
which maximize the monetary profit (or equivalently, minimize a cost
function) given a number of input variables and constraints. With
regard to energy arbitrage or ancillary services, this objective is to
maximize the profit Π from trading energy or power capacity over
a certain period of time. The decision variable is the power P or
an equivalent variable such as the current I measured at the battery
terminal. For different storage technologies, similar variables such as
a flow of water or heat may be chosen. The decision variable determines the amount of energy bought or sold during a particular time.
In this section, the transfer of energy from the ESS to the grid is
denoted *storage-to-grid* (S2G) as a generalization of the term V2G
since the ESS may not necessarily be a PEV or a BESS. The opposite
case is termed *grid-to-storage* (G2S). For ESSs other than batteries,
the term SOC therefore has to be understood as a generalized indicator describing the normalized energy content of the system. For
cases dealing with BESSs in particular, the abbreviations B2G and
grid-to-battery (G2B) will be used. As discussed in Section 2.3, prices
are generally constant over a certain period of time Δt which also applies to real-time markets. The problem can therefore be defined as

a discrete-time dynamic system where the discretization is given by the intervals in which prices are constant.

The profit π_k incurring in a period k can be expressed as

$$\pi_k = - \left[C_{\text{S2G},k} P_{\text{T},k} \Theta(1 - P_{\text{T},k}) + C_{\text{G2S},k} P_{\text{T},k} \Theta(P_{\text{T},k}) \right] \Delta t$$
$$- c_{\text{D}}(\vec{v}) \tag{4.1}$$

In this equation, $C_{\text{G2S},k}$ and $C_{\text{S2G},k}$ are the prices for buying and selling energy which are constant within a time interval k but which are generally assumed to vary across different periods. Prices are positive if purchasing a unit of energy costs money and negative in the opposite case. $P_{\text{T},k}$ denotes the power at the terminal of the storage system in period k, meaning that this is the value measured at the meter. Revenues are therefore defined as negative costs in Equation (4.1). The power is positive for charging and negative for discharging. The indicator function defined as

$$\Theta(P_{\text{T},k}) = \begin{cases} 1, & P_{\text{T},k} \geq 0 \quad \text{(G2S)} \\ 0, & P_{\text{T},k} < 0 \quad \text{(S2G)} \end{cases} \tag{4.2}$$

indicates that either the S2G or the G2S term is 0, depending on whether the battery is charged or discharged. The last term c_{D} denotes the system degradation cost which generally depends on a variety of parameters as indicated by the parameter vector $\vec{v}$. The corresponding cost function is generally non-linear so that the resulting optimization problem is also non-linear. The model developed to derive this cost function for the case of a BESS is presented in Section 4.5. There is furthermore an implicit cost which does not directly manifest itself in a cash flow. This is the energy loss due to charging/discharging efficiencies smaller than 1. The model developed to account for this cost is described in Sections 4.3 and 4.4.

Given a time series of N prices $\vec{C}_{\text{G2S}} = (C_{\text{G2S},1}, C_{\text{G2S},2}, ..., C_{\text{G2S},N})$ for G2S and another N prices $\vec{C}_{\text{S2G}} = (C_{\text{S2G},1}, C_{\text{S2G},2}, ..., C_{\text{S2G},N})$

for S2G, the scheduler attempts to compute a charging/discharging schedule $\vec{P} = (P_1, P_2, ..., P_N)$ which solves the following optimization problem:

$$\underset{\vec{P} \in \mathbb{R}^N}{\text{maximize}} \ \Pi(\vec{P}) = \sum_k \pi_k(P_k) \tag{4.3}$$

$$\text{subject to} \ P_{\min,k} \leq P_k \leq P_{\max,k} \tag{4.4}$$

$$\text{and} \ x_{\min,k} \leq x_k \leq x_{\max,k} \tag{4.5}$$

The control constraints $P_{\min,k}$ and $P_{\max,k}$ could be globally valid for all time periods if given by the maximum C-rate or the maximum connection power. They can also depend on the time period if they, for instance, define a constraint given by the fact that in certain periods the ESS might commit to delivering a particular power to an external load or if it needs to store energy generated by a PV module at a certain rate. The same holds for the state constraints $x_{\min,k}$ and $x_{\max,k}$ which can simply assume the values 0 and 1 if the entire SOC range is used. Depending on the use case, however, different values may apply. For instance, in the case of a PEV, the driver may not wish the SOC to go below a certain threshold at certain times which in this case would be defined by setting the value of $x_{\min,k}$ to an appropriate value for the respective period.

Assuming the price information is updated at each time step, the profit optimization can be treated as a *rolling horizon* procedure, also commonly termed *receding horizon* procedure in operations research or *model predictive control* in engineering control theory. In this case, if the current time period is period i, the problem specified in Equation (4.3) is solved for a sequence of periods from i to $i + l$ with l denoting the lookahead. From the computed schedule, only the solution for period i is taken. In the subsequent period, the optimization is then performed for the periods $i + 1$ to $i + l + 1$ by taking advantage of the newly available information for period $i + l + 1$ as well as potentially updated information for the periods up to $i + l$.

4.2.1 Formulation as Dynamic Programming Problem

As the setting exhibits the characteristics of a sequential decision-making problem, a suitable approach to solve the problem can be realized by means of a DP formulation [156]. The main advantage of a DP algorithm is that it can provide a globally optimal solution for a generally non-linear process model with non-linear equality or inequality constraints [157]. A DP algorithm can furthermore easily be parallelized since it involves computations which are independent from each other, thus allowing to effectively use HPC resources. The computational complexity with regard to the lookahead is $O(n)$ which is an advantage compared to a *mixed integer non-linear program* (MINLP) formulation which is NP-hard.

In accordance with the DP formalism the system is described by the temporal development of its state through a transition function

$$x_{k+1} = f_k(x_k, P_k), \quad k = 0, 1, ..., N - 1 \tag{4.6}$$

The state x_k which in the case of an ESS corresponds to the SOC is an element of the discretized state space S_k which, in accordance with Restriction (4.5), is given by

$$S_k = \{x_k \in \mathbb{R} \,|\, 0 \leq x_k \leq 1\} \tag{4.7}$$

The control variable P_k belongs to the control space

$$C_k = \{P_k \in \mathbb{R} \,|\, P_{\mathrm{min},k} \leq P_k \leq P_{\mathrm{max},k}\} \tag{4.8}$$

which is in accordance with Restriction (4.4).

Possible values for P_k are further constrained to a subset $U(x_k) \subset C_k$ which depends upon the current state of the system x_k so that $P_k \in U_k(x_k)$ for all k and $x_k \in S_k$. As a result of Constraint (4.7), this subset can be written as

$$U_k(x_k) = \{P_k \in \mathbb{R} \,|\, 0 \leq f_k(x_k, P_k) \leq 1\} \tag{4.9}$$

The class of control policies can be expressed in the form

$$\gamma = \{\mu_0, ..., \mu_{N-1}\} \tag{4.10}$$

where μ_k maps the states x_k into controls $P_k = \mu_k(x_k)$ so that $\mu_k(x_k) \in U_k(x_k)$ for all $x_k \in S_k$.

For given cost functions g_k with $k = 0, 1, ..., N$, the total cost J of a control strategy γ which starts at a state x_0 is

$$J_\gamma(x_0) = g_N(x_N) + \sum_{k=0}^{N-1} g_k(x_k, \mu_k(x_k)) \tag{4.11}$$

The cost function g_k in this case is simply defined by $-\pi_k$ as given by Equation (4.1). The cost of the final step g_N is set to the negative value of the energy content contained in the ESS, valued at the average selling price at the respective period. An optimal control strategy $\gamma*$ minimizes these costs so that

$$J_{\gamma*}(x_0) = \min_{\gamma \in \Psi} J_\gamma(x_0) \tag{4.12}$$

with Ψ representing the set of all admissible policies.

4.2.2 Solution Algorithm

To solve the described problem, the SOC is discretized into M steps so that the state can be written as x_k^j with $j = 0, 1, ..., M-1$. An optimal control strategy is then computed by applying a DP algorithm. The algorithm starts from the last state of the problem of which the cost is set to

$$J_N^j(x_N^j) = g_N^j(x_N^j) = -x_N^j E_{\text{ESS}} \bar{C} \tag{4.13}$$

In this equation, $\bar{C}$ denotes the average price at which the energy could be sold and E_{ESS} the capacity of the ESS in energy units. Subsequently, proceeding backwards in time, the cost g_{N-1}^j leading to the total cost $J(x_{N-1}^j)$ for every j is calculated. The transition between

$N - 1$ and N leading to the lowest total cost for each j then yields the desired $P_{N-1}^j = \mu_{N-1}^j(x_{N-1}^j)$. Continuing this procedure up to the first stage $k = 0$ leads to the minimized costs $J_{\gamma^k_*}^j(x_0^j)$ and an optimal control policy $\gamma^j* = \left\{\mu_0^j*, ..., \mu_{N-1}^j*\right\}$ for every j corresponding to x_0^j. Since the initial x_0 is given, j corresponding to this value is chosen in order to identify the applicable control policy. Presuming deterministic prices, this approach can be proven to find a global optimum for the described problem [156].

4.3 Equivalent Circuit Model

As discussed in Section 2.4, Li-ion batteries have been emerging as the leading technology for small-scale BESS and are now the state-of-the-art technology for PEV battery packs. As the applications under consideration in this work mainly concern the grid integration of PEVs and in a greater context residential-scale stationary batteries, the models presented in this chapter are targeted at Li-ion batteries.

There are three main families of battery modeling techniques including mathematical models, electrochemical models and EC models. Mathematical models are analytical or stochastic abstractions which aim to describe the behavior of a battery cell with a small set of equations. These models can therefore be considered a rather coarse approach to battery modeling. On the other side of the spectrum are electrochemical models which attempt to describe the battery at a very fundamental level. A compromise between complexity and accuracy is given by EC models which abstractly describe a cell's electrochemical behavior by means of electrical components. A comprehensive overview of models corresponding to the three different categories in the context of PEVs is given in [158, 159].

In this work, the main aspects of interest are relationships between SOC changes and terminal power as well as the cell's voltage characteristic. Considering the trade-off between model accuracy and

Table 4.1: Technical specifications of the reference cell Sanyo UR18650E and the test cell Panasonic UR18650RX.

	Sanyo UR18650E (reference cell)	Panasonic UR18650RX (test cell)
Typical capacity [Ah]	2.15	2.05
Typical voltage [V]	3.6	3.6
Min./max. voltage [V]	2.5/4.2	2.5/4.2

computational complexity, the class of EC models can be used to address these requirements. This compromise is therefore chosen for the model employed in this work and is thus subject of this section.

As discussed in great detail in [159–161], there is a variety of EC model types available which may be chosen according to cell types and application scenarios. In this work, an extension of a single resistance model is used with parametrizations based on experimental measurements as discussed in the remainder of this section.

4.3.1 Reference Cell and Test Cell

Since battery properties greatly vary among different types of Li-ion batteries, the parametrizations of the models presented in this chapter are based on a particular reference cell. The battery degradation model parametrization presented in Section 4.5.7 is based on the Sanyo UR18650E, a cell with an 18650 form factor. This cell's cathode active material is composed of $Li(NiMnCo)O_2$ (NMC) and the anode consists of graphite. At the time of the experimental determination of the parameters of the EC model presented in Section 4.3, the same cell was not on the market anymore which is why a closely related successor model was used. The basic specifications of the Sanyo UR18650E and this substitute cell of the type Panasonic UR18650RX are listed in Table 4.1.

4.3.2 Open Circuit Voltage

The starting point of the discussion is the voltage

$$U_T = U_{OC} + U_R + U_{pol} \tag{4.14}$$

$$= U_{OC} + R_i I + U_{pol} \tag{4.15}$$

U_T denotes the voltage measured at the battery's terminal, U_{OC} the *open circuit voltage* (OCV) and U_R the voltage drop at the internal resistance R_i which depends on the current I. Finally, U_{pol} denotes the polarization voltage which is a result of non-ohmic resistance.

The OCV corresponds to the battery's equilibrium potential which varies across the cell's SOC. A functional representation of this OCV characteristic depending on SOC is required by the battery degradation model as discussed in Section 4.5. The OCV measurement for determining this characteristic was conducted with two similar cells under temperature controlled conditions at 35°C. Initially, cells were prepared by discharging up to a voltage of 2.75 V, followed by a relaxation period of 30 min. Afterwards, a charging process was conducted up to a cut-off voltage of 4.2 V. In SOC steps of 0.1 the charging/discharging was interrupted for voltage relaxation. As the characteristic of the voltage relaxation shows that after a time of 20 min the further voltage change is negligibly low, the value at this point was taken as the OCV. This measurement was conducted for different C-rates in the range from 0.1 C to 0.7 C.

The measurement results are shown in Figure 4.1a. As there is no dependency on the applied C-rate, the figure does not differentiate between the various parameter settings. The continuous shape of the curve shows that the variability among different measurements is very small. The measurement points do not appear at multiple SOC values of 0.1 since different C-rates result in different total charged Ah. As a result, the discharging process does not always start at a fully charged cell according to the measured capacity. Ah-based

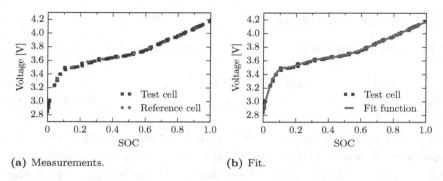

(a) Measurements.   (b) Fit.

Figure 4.1: Measured OCV data and fit function.

discharging therefore does not necessarily hit SOC values which are multiples of 0.1.

The OCV curve of the reference cell as provided in [162] starts at a voltage of 3.32 V, indicating that the cell was not discharged as deeply as in the given measurement. This voltage corresponds to an SOC of 0.056 of the test cell. Re-scaling the data from [162] to an SOC range from 0.056 to 1 shows a very good accordance of both cell types as indicated in Figure 4.1.

The functional representation of the cell's OCV characteristic is obtained by fitting the following sum of sigmoid functions to the measured data.

$$
\begin{aligned}
U_{OC}(x) =& K_0 + K_1 \frac{1}{1 + e^{\alpha_1(x-\beta_1)}} + K_2 \frac{1}{1 + e^{\alpha_2(x-\beta_2)}} \\
& + K_3 \frac{1}{1 + e^{\alpha_3(x-1)}} + K_4 \frac{1}{1 + e^{\alpha_4 x}} + K_5 x \qquad (4.16)
\end{aligned}
$$

This formulation adopted from [163] yields a better representation of the characteristic voltage plateaus than a simple polynomial fit. The fit parameters are provided in Table 4.2 and the corresponding fit to the data points is shown in Figure 4.1b.

Table 4.2: OCV fit parameters.

Param.	Value	Param.	Value	Param.	Value
K_0	-433.861	α_1	2.702	β_1	0.0015
K_1	154.306	α_2	28.154	β_2	0.116
K_2	0.558	α_3	1.601		
K_3	284.139	α_4	-1.534		
K_4	244.892				
K_5	84.789				

4.3.3 Internal Resistance

The OCV does not accurately represent the cell's voltage under load or during charging. This is because the cell's internal resistance results in a voltage drop which causes the terminal voltage to be lower than the OCV while discharging and requires a voltage higher than the OCV during charging. The related losses consist of an ohmic component proportional to the current (*ohmic drop* or *ohmic polarization*) as well as non-ohmic components. Ohmic resistance consists of the electronic resistances of the electrodes, current collectors and contacts, ionic resistance of the electrolyte and contact resistance between current collectors and active material. In turn, the non-ohmic component is composed of activation and concentration polarization. Concentration polarization results from the build-up of concentration gradients due to finite diffusion velocities which lead to depletion of active material in the vicinity of the electrodes. It therefore typically increases with current. Activation polarization is a kinetic limitation resulting from the energy which is required to transfer electrons between the electrodes and the electrolyte. For both ohmic and non-ohmic resistance, simple parametrizations are established in this section.

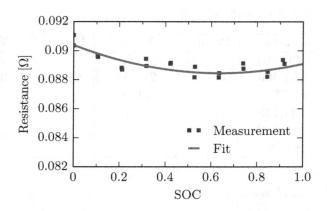

Figure 4.2: Illustration of ohmic resistance.

4.3.3.1 Ohmic Resistance

The ohmic charge/discharge resistances were measured in SOC steps of 0.1 at different C-rates ranging from 0.1 C to 0.7 C. As expected for a Li-ion cell, the ohmic resistance is fairly constant over the entire SOC range with slightly higher values at high and low SOCs (about 1 % deviation from the mean in the case of very high/low SOCs). An illustration of the shape of the curve is given in Figure 4.2. In a simple approximation, this can be represented by a second order polynomial

$$R(x) = a_0 x^2 + a_1 x + a_2 \tag{4.17}$$

which accounts for the higher resistance values at low and high SOCs. The parameters of a least squares fit to the measured data are given in Table 4.3. In this particular case typical for a Li-ion battery, the resistance's SOC dependency is very small so that it could, in principle, be replaced by its average value. For the sake of generality, however, the SOC dependency is preserved in the model.

Table 4.3: Fit parameters and mean values of the internal ohmic resistance.

Parameter	Value
a_0	0.00481
a_1	-0.00612
a_2	0.09314
$\overline{R}$	0.09169

4.3.3.2 Non-Ohmic Resistance

The non-ohmic component of the internal resistance causes an additional voltage drop. Here it is expressed through the polarization voltage which is considered as the difference between the measured charge voltage U and the OCV component as well as the ohmic IR drop $U_{OCV} + R_i I$. As shown in Figure 4.3, the polarization voltage during charging strongly depends on the SOC in the range of low SOCs between 0 and 0.1 and then remains relatively constant until the cell is fully charged. It furthermore slightly depends upon the C-rate with higher C-rates implying higher voltages. The part of the voltage curve for $x \leq 0.1$ is approximated by a function consisting of an exponential term, a polynomial term and a constant as follows:

$$U_{pol} = a \exp(bx) + cx^d + e \quad (x \leq 0.1) \tag{4.18}$$
$$U_{pol} = f \quad (x > 0.1) \tag{4.19}$$

As the voltage in the region with $x > 0.1$ is almost independent of the SOC, in this region it is approximated by a constant f. The same holds for the voltage during discharging which is also approximated by the constant f. The parameters resulting from a least squares fit are listed in Table 4.4.

The results indicate that in the considered range of C-rates the polarization voltage increases roughly linearly with the current. The increase factor is therefore computed using a linear model which leads

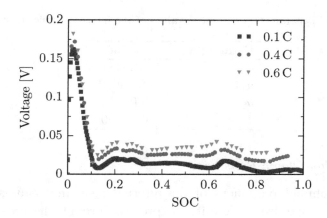

Figure 4.3: Charge polarization voltage of the test cell.

to an offset factor of $\Delta U(I) = 0.04\,\text{V}$ per $1\,\text{C}$ for $x > 0.1$. For $x \leq 0.1$, the factor has a slight but insignificant SOC dependency between 0.02 and 0.1. Below 0.02 it decreases to 0 which is approximated by a linear function with slope 2.

To compute the voltage curve depending on the C-rate, the offset parameter is applied to the fit for 0.1 C. A comparison between measured terminal voltages and the corresponding model output is shown in Figure 4.4. The measured data results from a different measurement than the one used to obtain the fit parameters. It can be seen that the functional dependency described above represents the data

Table 4.4: Fit parameters for Equations (4.18) and (4.19) for 0.1 C.

Parameter	Value
a	-0.201232
b	-142.46578
c	-1.98023
d	1
e	0.2138
f	0.01479

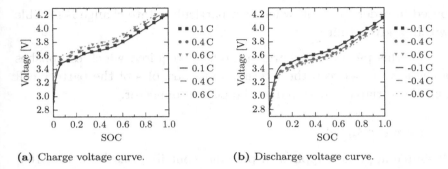

(a) Charge voltage curve.   **(b)** Discharge voltage curve.

Figure 4.4: Comparison between measured data and fit functions at 0.1 C,
0.4 C and 0.6 C. Measured data is indicated by markers, the voltages computed by the model are shown as continuous, dashed and
dotted lines.

well for both charging and discharging processes in the investigated
current range. To validate the statement for significantly smaller or
larger currents, additional measurements would be required. For the
investigations presented in Chapter 5, the C-rate is therefore limited
to 1. As also discussed in Section 4.6, the phenomenological description presented here cannot be regarded a stringent model of a li-ion
battery cell but is considered sufficient given this work's primary focus on a charge scheduling framework.

4.4 Charging Model

The scheduling algorithm needs to commit to a certain terminal power
P_T at a particular time or to a specific amount of energy ΔE which
is charged or delivered during a certain time period Δt. How much
power or energy can be provided or consumed depends on the cell's
internal state, most importantly its SOC x. It is therefore necessary
to know what terminal power corresponds to a certain incremental
SOC change dx/dt. Similarly, it has to be known what amount of energy ΔE corresponds to an SOC change Δx. Furthermore, in some
cases, the time t required to reach a specific SOC needs to be com-

puted in order to assess whether a particular state change is feasible within a given time.

For this purpose, a charging model is required which establishes a connection between the internal state variables of the battery cell and the variables measured at the power connector.

4.4.1 Linear Model

A simple approach applied in previous contributions [109, 110, 164] is to establish a linear relationship between SOC change and power. The SOC x at time t is given by the start SOC x_0 and the SOC change so that

$$x(t) = x_0 + \Delta x_{\text{G2B/B2G}}(t) \tag{4.20}$$

with the linear relationship between SOC change and power

$$\Delta x_{\text{G2B}}(t) = \frac{P_{\text{T}}t}{E_{\text{ESS}}}\eta_{\text{G2B}} \tag{4.21}$$

$$\Delta x_{\text{B2G}}(t) = \frac{P_{\text{T}}t}{E_{\text{ESS}}}\frac{1}{\eta_{\text{B2G}}} \tag{4.22}$$

Here, η_{G2B} and η_{B2G} denote the charge and discharge efficiencies, respectively and E_{ESS} represents the battery capacity in energy units. The boundary conditions in this case are given by

$$P_{\text{min}} \leq P_{\text{T}} \leq P_{\text{max}} \tag{4.23}$$

and

$$0 \leq x \leq 1 \tag{4.24}$$

4.4.2 Equivalent Circuit Model-Based Charging Model

The simplified approach described in Section 4.4.1 ignores a number of aspects which depending on the application can be relevant in reality. First, a cell is generally not charged or discharged at a constant power over the entire SOC range. Instead, different charging modes such as

CC and CV charging have to be distinguished. As the terms indicate, CC charging implies a constant current while the voltage is subject to change as opposed to CV charging where voltage is kept constant while the current changes. Furthermore, charge/discharge efficiencies cannot generally be assumed to be constant over the entire SOC range and also depend on the applied current. A better approach therefore builds on an EC model as presented in Section 4.3. This informs about the cell's voltage as a function of SOC and current which allows accounting for the different modes of a charging process. In contrast to the constant efficiency factor η, in this case the efficiency which is generally dependent on SOC and current is implicitly considered by means of the cell's internal resistance.

4.4.2.1 Basic Equations

The state of the cell is described by its SOC x which is defined as

$$x = \frac{q}{Q_0} \tag{4.25}$$

In this relation, q denotes the charge contained in the cell and Q_0 the cell's capacity, both given in Ah. The SOC change with time is therefore given by

$$\frac{\mathrm{d}x}{\mathrm{d}t} = \frac{1}{Q_0}\frac{\mathrm{d}q}{\mathrm{d}t} \tag{4.26}$$

With the current

$$I = \frac{\mathrm{d}q}{\mathrm{d}t} \tag{4.27}$$

this leads to

$$x(t) = \frac{1}{Q_0} \int I(t)\,\mathrm{d}t \tag{4.28}$$

which for a period Δt during which the current is constant turns into

$$x(t) = x_0 + \frac{I\Delta t}{Q_0} \tag{4.29}$$

with x_0 denoting the SOC at which the cycling begins. The power at the cell's terminal is given by

$$P_T(t) = U_T(t)I(t) \tag{4.30}$$

Consequently, the energy transferred over time is

$$E_T(t) = \int U_T(t)I(t)\,\mathrm{d}t \tag{4.31}$$

4.4.2.2 Constant Current/Constant Voltage Mode Charging

The presented methodology considers both CC and CV charging. A cell is typically charged in CC mode until a cut-off voltage U' is reached at a charge level Q'. (In this section, all variables with a prime symbol indicate the point of the transition from CC to CV mode). To avoid battery damage, the charger then keeps the voltage constant while linearly decreasing the current from a starting value I' over the charge q. After the current has dropped below a certain value, e.g., 5 % of the rated charge current, the cell is considered fully charged. According to the voltage model in Equation (4.14), the point at which the CV phase is entered depends upon the C-rate with higher C-rates leading to an earlier onset of CV charging.

Current

The currents in the CC and CV phase can be written as a function of q in the form

$$I(q) = \mathrm{const} \qquad\qquad\qquad U < U' \tag{4.32}$$

$$I(q) = I' - \frac{I'}{Q_0 - Q'}(q - Q') \quad U \geq U' \tag{4.33}$$

These equations are also needed as a function of time. In CC mode, this is trivial because the current is given by

$$I(t) = I(q) = \text{const} \quad U < U' \tag{4.34}$$

For CV charging, Equation (4.33) can be re-formulated as a function of time using Equation (4.28). Differentiating Equation (4.33) with respect to t leads to the differential equation

$$\frac{\mathrm{d}I(q)}{\mathrm{d}t} = -\frac{I'}{Q_0 - Q'}\frac{\mathrm{d}q}{\mathrm{d}t} \tag{4.35}$$

$$= -\frac{I'}{Q_0 - Q'}I(t) \tag{4.36}$$

Using an *exponential ansatz* $I(t) = a\exp(\lambda t)$ with the boundary condition $I(t = 0) = I'$, this turns into

$$I(t) = I'\exp\left(-\frac{I'}{Q_0 - Q'}t\right) \quad U \geq U' \tag{4.37}$$

Energy Transfer

Equation (4.31) describes the energy transfer as a function of time. The scheduling algorithm, however, computes the amount of energy transferred between SOC states rather than time intervals because the state space of the scheduling approach consists of discrete SOC steps. Furthermore, the voltage U_T as defined in Equation (4.14) is given as a function of SOC only. The time dependency in Equation (4.31) therefore needs to be translated into an SOC dependency. This can be done in separate ways for the CC and CV mode.

In CC mode, $I(t) = I(q) = \text{const}$ so that Equation (4.29) can be written as

$$x(t) = x_0 + \frac{It}{Q_0} \tag{4.38}$$

This can be differentiated with respect to t leading to

$$\frac{\mathrm{d}x}{\mathrm{d}t} = \frac{I}{Q_0} \tag{4.39}$$

so that

$$\mathrm{d}t = \frac{Q_0}{I}\,\mathrm{d}x \tag{4.40}$$

Using this substitution, the integral in Equation (4.31) can be written as

$$E_{\mathrm{T}}^{\mathrm{CC}}(x) = Q_0 \int U(x)\,\mathrm{d}x \quad U < U' \tag{4.41}$$

For transforming the integral limits, t in Equation (4.31) is substituted by Equation (4.38) solved for t.

In CV mode, $U(t) = U' = $ const so that only the number of charges transferred at this voltage needs to be considered. With Equation (4.27), Equation (4.31) can therefore be written as

$$E_{\mathrm{T}}^{\mathrm{CV}}(x) = U' \int \mathrm{d}x \quad U \geq U' \tag{4.42}$$

4.4.2.3 Feasibility Conditions

The system design requires the scheduling algorithm to guarantee that a certain amount of energy ΔE can be purchased or dispatched during a period of time Δt. In order to find an optimal schedule, the algorithm therefore needs to assess for any SOC transition (x_m, x_n) whether it can be performed during the time Δt and how much charged/dispatched energy it corresponds to.

Dispatching is assumed to be conducted in a CC mode, charging may be performed in CC or CV mode, depending on the actual voltage. Since this does not guarantee a constant power, it is the task of the aggregator to balance power deviations across different BESSs. By making the time Δt arbitrarily short, the deviation of this approach from a constant power commitment can, however, be made negligibly small at the expense of a higher computational cost.

The constraints for the solver are given by the state variable x and the control variable I through

$$x_{\min} \leq x \leq x_{\max} \tag{4.43}$$

$$I_{\min} \leq I \leq I_{\max} \tag{4.44}$$

where $I_{\min}$ and $I_{\max}$ denote the maximum discharge and charge currents, respectively. Equation (4.43) can simply be fulfilled by limiting the state space for the stage under consideration to the indicated range. For the control constraint in Equation (4.44), the charging model of the cell has to be used.

In CC mode, the required current corresponding to a state transition (x_m, x_n) is given by

$$I = Q_0(x_n - x_m)t^{-1} \tag{4.45}$$

which can be checked against Equation (4.44). This check is more complicated if at least a part of the charging process lies within the CV domain. The state transition (x_m, x_n) corresponds to a charge throughput $\Delta q_{m,n} = Q_0(x_n - x_m)$. According to Equation (4.27), this charge throughput is related to the current through

$$\Delta q_{m,n} = I^{CC} \Delta t^{CC} + \int_0^{\Delta t - \Delta t^{CC}} I^{CV}(t) \, \mathrm{d}x \tag{4.46}$$

with the constant current I^{CC} in CC mode, the time $\Delta t^{CC} = (Q' - Q_0 x_m) I^{CC}$ spent in CC mode and the current $I^{CV}(t)$ in CV mode given by Equation (4.37). Q' follows from the voltage model in Equation (4.14) with $U_T(I^{CC}) = U'$. After integration, Equation (4.46) can be numerically solved for the smallest possible I^{CC} which is the starting current required to reach the targeted SOC within time Δt.

If the charging process starts in the CV domain, the first term in Equation (4.46) equals 0. If no solution is found, the state transition is not possible in the given period of time. Similarly, if the computed I^{CC} does not match the condition according to Equation (4.44), the solution is infeasible.

4.5 Battery Degradation Monetization Model

Ideally, battery operation should only result in reversible changes in the phases at the electrodes and in modifications of their physical properties. There is, however, a number of undesirable side effects which lead to progressing cell degradation resulting in a finite lifetime. For cost-optimal scheduling, it is crucial to take these processes into account in order to avoid excessive battery degradation. The first step for considering battery degradation for optimal scheduling is to quantify the degradation process physically, i.e., to understand how much physical damage the battery cell experiences under particular operating conditions. At a high level, this damage can be measured by the change of a physical quantity such as the capacity or the internal resistance. This change does not only depend on the present operating conditions but is also dependent on how much aging the cell has undergone already. In a next step, the incurred physical damage needs to be translated into monetary units so that it can be considered as part of the cost function of the optimization problem. This section presents an approach addressing both of these steps.

4.5.1 Battery Degradation Processes

As discussed in a publication prepared in the course of this work [109], the degradation of a cell manifests itself in the loss of capacity (measured in Ah) as well as in an increase in impedance which cause energy fade (measured in Wh) and power fade (measured in W). While the loss of capacity reduces the amount of electrical energy which can be stored, impedance increase increases ohmic losses thus negatively affecting the cell's charging/discharging efficiency.

Cell degradation is primarily a result of decreasing negative electrode performance as well as of the loss of cyclable lithium. This is related to the ongoing growth of the *solid electrolyte interface* (SEI) at the graphite anode [165–167] as well as to lithium plating at high charging currents and low temperatures. When the cell is cycled at large DODs, the electrodes undergo volumetric changes during lithium intercalation resulting in mechanical stress. This may result in loss of contact in the electrode or the current collector [166, 168, 169] or lead to the formation of micro cracks [50, 167]. These cracks result in the exposure of fresh graphite surfaces to the electrolyte, thus leading to further SEI formation and electrolyte consumption [50, 165, 166, 170]. In addition to capacity fade, these mechanisms are closely related to the increase of cell impedance. The ongoing SEI reformation causes a constantly growing surface layer with a low conductivity and low diffusivity, causing an increase in the charge transfer resistance [170]. The loss of active material leads to higher local currents and local SOC variations which in turn accelerate the degradation process [171].

In general, degradation can be caused by calendar aging or cycle aging. The different aging mechanisms depend on environmental and operating conditions including the cell's temperature, the DOD, the charge and discharge current, as well as the used SOC range [168, 170, 172, 173]. Calendar aging primarily depends on the storage temperature [174] as well as the SOC [175]. This is because high temperatures increase the velocity of side reactions such as metal dissolution which increases the rate of capacity loss [176]. As the SOC represents the quantity of lithium ions present in the negative electrode, a high SOC increases the potential for losing large amounts of cyclable lithium on the anode/electrolyte interface as a consequence of SEI growth. Elevated SOCs therefore result in faster calendar aging [176]. In turn, cycle aging occurs while the cell is in operation. Influencing factors in this case are especially C-rate, initial SOC and SOC swing Δx. High C-rates lead to higher losses in terms of *joule heating* which further leads to an increased amount and speed of parasitic side reactions [177]. Cycling at extreme C-rates close to a fully charged or discharged cell accelerates the increase of the charge-transfer resistance as shown in [178]. The primary reason for high battery wear at extreme SOCs can be inferred from the characteristic of Li-ion batteries that the internal resistance is usually higher at both ends of the SOC range which results in higher joule heating [133]. Similarly to high C-rates, high Δx values result in an accelerated loss of battery power and capacity over time [179–181] which has been partly attributed to structural changes and a phase transition of the cathode material [166, 176].

Various studies have been performed to understand these mechanisms and to establish a quantitative relation between these aging effects and the corresponding control parameters [173, 182]. The temperature dependency of calendar aging can generally be described using an Arrhenius equation as has been extensively discussed in the literature [183–187]. At the same time, the lithium loss due to SEI formation and thus calendar capacity loss is usually considered to be proportional to the square root of time [179, 185, 188, 189]. There is, however, other empirical work stating that a dependency of capacity loss $\propto t^{0.75}$ can be observed [190]. The SOC dependency of aging is reported to be exponential in [186, 191, 192]. Cycle aging has been modeled as a cubic polynomial of Δx in [112] and with a power law for the total number of cycles depending on Δx in [110, 131, 132, 193]. For an in-depth discussion of aging mechanisms in Li-ion batteries, the reader is referred to [50, 166, 176].

4.5.2 Basic Model Assumptions

The general considerations on battery aging processes are based on a number of presumptions which the model is built on:

1. The processes leading to calendar and cycle aging are independent from each other as supported by the presentation in [190]. This allows treating the calendar and cycle components of the capacity and resistance changes separately. This means that calendar and cycle capacity losses as well as calendar and cycle resistance increases can be linearly superpositioned as follows:

$$\Delta\Gamma_{tot} = \Delta\Gamma_{cal} + \Delta\Gamma_{cyc} \tag{4.47}$$

$$\Delta R_{tot} = \Delta R_{cal} + \Delta R_{cyc} \tag{4.48}$$

2. The primary drivers for calendar aging are the temperature T and the SOC x. For a particular battery, the SOC can be represented by the OCV U_{OCV}. In case a battery is cycled between two SOCs, x_m and x_n, the voltage changes. It is therefore assumed that the average voltage

$$\overline{U}(x_m, x_n) = \frac{U(x_m) + U(x_n)}{2} \tag{4.49}$$

$$= \frac{U(x_m) + U(x_m + \Delta x)}{2} \tag{4.50}$$

is a good approximation for the voltage determining calendar aging. This is a simplification since the voltage curves during charging and discharging are not exactly identical as a consequence of the IR drop. As this leads to a slight voltage underestimation in the case of charging and a slight overestimation in the opposite case, the small errors compensate each other so that the relevance of the inaccuracy can be considered negligible. The calendar capacity and resistance change can therefore be written as

$$\Delta\Gamma_{cal} = \Delta\Gamma_{cal}(T, \overline{U}) \tag{4.51}$$

$$\Delta R_{cal} = \Delta R_{cal}(T, \overline{U}) \tag{4.52}$$

3. The primary parameters which determine cycle aging are the SOC swing $\Delta x = |x_n - x_m|$ and the absolute SOC at which the cycling occurs. The latter can again be represented by the average voltage $\overline{U}(x_m, x_n)$. The cycle capacity and resistance change can therefore be expressed as

$$\Delta\Gamma_{\text{cyc}} = \Delta\Gamma_{\text{cyc}}(\Delta x, \overline{U}) \tag{4.53}$$

$$\Delta R_{\text{cyc}} = \Delta R_{\text{cyc}}(\Delta x, \overline{U}) \tag{4.54}$$

4.5.3 Physical Degradation Formalization

At any point in time, the scheduling mechanism requires an estimate of the damage which will incur to the battery in the next time period, i.e., how much capacity will be lost or how much resistance increase has to be expected. In accordance with the formulation in Assumption 1, capacity and internal resistance are treated independently from each other. As the following considerations are the same for both capacity and internal resistance, the symbol Φ is used to denote either one of both quantities. By definition, Φ starts with a value of 1 at the cell's *beginning of life* (BOL). As capacity decreases during degradation while the opposite is the case for internal resistance, its value decreases over time if it represents capacity while it increases if it represents internal resistance.

The information on how much degradation occurs under uniform cycling conditions is generally experimentally determined using cycling tests by operating multiple sets of batteries under different conditions while repeatedly measuring the value of Φ. A schematic example on the information which may be provided by a cycling test if cells are cycled under two different cycling regimes $\vec{v}_1 = (T_1, \Delta x_1, \overline{U}_1)$ and $\vec{v}_2 = (T_2, \Delta x_2, \overline{U}_2)$ is shown in Figure 4.5a. Without loss of generality, the considerations in this example are performed for the cell's capacity Γ. Assuming a cell was operated under any of the regimes $\vec{v}_1$ or $\vec{v}_2$ since the BOL, the change of Γ could be taken from these graphs

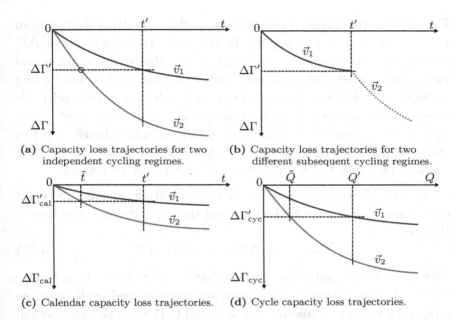

(a) Capacity loss trajectories for two independent cycling regimes.

(b) Capacity loss trajectories for two different subsequent cycling regimes.

(c) Calendar capacity loss trajectories.

(d) Cycle capacity loss trajectories.

Figure 4.5: Illustration of calendar and cycle capacity loss for two different cycling regimes $\vec{v}_1$ and $\vec{v}_2$.

for any point t. In reality, however, as illustrated in Figure 4.5b, operating conditions are not uniform so that at a time $t \leq t'$ the cell might be operated at a regime $\vec{v}_1$ followed by a regime $\vec{v}_2$ for $t > t'$. In the example in Figure 4.5b, at t' a certain amount of energy has been cycled through the battery and a certain time has passed. This has resulted in cycle and calendar aging leading to a capacity loss $\Delta\Gamma'$ as follows from the trajectory corresponding to $\vec{v}_1$. From Figure 4.5a it can, however, not be concluded how the trajectory indicated by the dashed line continues after t'.

Capacity is a physical quantity which has to be continuous at any point including the one at t'. A conceivable but incorrect option would therefore be to follow the dashed line which is the trajectory corresponding to $\vec{v}_2$ starting from the point at $\Delta\Gamma'$ as indicated by the circle in Figure 4.5a. In the hypothetical aging test corresponding to

Figure 4.5a, this point, however, was reached in a time shorter than t', e.g., due to heavier cycling. In this case, the capacity loss $\Delta\Gamma'$ related to $\vec{v}_2$ is therefore to a larger degree due to cycle aging and to a lesser degree due to calendar aging as compared to the cycling regime $\vec{v}_1$. Following Assumption 1 that calendar aging and cycle aging are physically independent processes, the cell operated under $\vec{v}_1$ and the cell operated under $\vec{v}_2$ in Figure 4.5a are therefore not in the same physical state after losing a total $\Delta\Gamma'$ of their capacity. The dashed trajectory in Figure 4.5b can therefore not be correct.

Finding the correct trajectory requires treating calendar and cycle degradation separately by formulating the capacity as a function of time as

$$\Gamma(t) = 1 + \Delta\Gamma_{\text{cal}}(t) + \Delta\Gamma_{\text{cyc}}(t) \qquad (4.55)$$

with the calendar and cycle capacity changes $\Delta\Gamma_{\text{cal}}$ and $\Delta\Gamma_{\text{cyc}}$, respectively. A degradation study separating calendar and cycle aging as required for this formulation can be found in [190]. The hypothetical calendar aging trajectories for $\vec{v}_1$ and $\vec{v}_2$ are shown in Figure 4.5c and the cycle aging trajectories for $\vec{v}_1$ and $\vec{v}_2$ are depicted in Figure 4.5d. The calendar aging trajectory is then a function of time t while the cycle aging trajectory is a function of charge throughput Q. At any point in time including the one at t', both functions describing the calendar and cycle capacity loss have to be continuous. As $\Delta\Gamma_{\text{cal}}$ is a function of t while $\Delta\Gamma_{\text{cyc}}$ is a function of Q, the first one has to be a continuous function in t and the latter a continuous function in Q. By denoting the function describing the calendar capacity decrease corresponding to $\vec{v}_1$ and $\vec{v}_2$ as $\Delta\Gamma_{1,\text{cal}}(\vec{v}_1, t)$ and $\Delta\Gamma_{2,\text{cal}}(\vec{v}_2, t)$, respectively and the function describing the cycle capacity decrease as $\Delta\Gamma_{1,\text{cyc}}(\vec{v}_1, Q)$ and $\Delta\Gamma_{2,\text{cyc}}(\vec{v}_2, Q)$, respectively, this means that

$$\Delta\Gamma_{1,\text{cal}}(\vec{v}_1, t)|_{t'} = \Delta\Gamma_{2,\text{cal}}(\vec{v}_2, t)|_{\tilde{t}} \qquad (4.56)$$

$$\Delta\Gamma_{1,\text{cyc}}(\vec{v}_1, Q)|_{Q'} = \Delta\Gamma_{2,\text{cyc}}(\vec{v}_2, Q)|_{\tilde{Q}} \qquad (4.57)$$

In these equations, $\tilde{t}$ and $\tilde{Q}$ are the reference time and charge which would have accumulated if the cell had reached its state at t' under the operating conditions $\vec{v}_2$. These reference points can be computed through the inverse functions

$$\tilde{t} = \Delta\Gamma_{2,\text{cal}}^{-1}(\Delta\Gamma_1) \tag{4.58}$$

$$\tilde{Q} = \Delta\Gamma_{2,\text{cyc}}^{-1}(\Delta\Gamma_1) \tag{4.59}$$

They define the points at the aging trajectory corresponding to $\vec{v}_2$ where the degradation process proceeds at t'.

As previously mentioned, these considerations are not restricted to the capacity so that the presented equations can also be written in their general form for Φ with

$$\Phi(t) = 1 + \Delta\Phi_{\text{cal}}(t) + \Delta\Phi_{\text{cyc}}(t) \tag{4.60}$$

denoting the change of the physical quantity under consideration. With this modification and by generalizing $\vec{v}_1$ and $\vec{v}_2$ to $\vec{v}_i$ and $\vec{v}_{i+1}$, the continuity condition is expressed as

$$\Delta\Phi_{i,\text{cal}}(\vec{v}_i, t)|_{t'} = \Delta\Phi_{i+1,\text{cal}}(\vec{v}_{i+1}, t)|_{\tilde{t}} \tag{4.61}$$

$$\Delta\Phi_{i,\text{cyc}}(\vec{v}_i, Q)|_{Q'} = \Delta\Phi_{i+1,\text{cyc}}(\vec{v}_{i+1}, Q)|_{\tilde{Q}} \tag{4.62}$$

and the representation for the reference points as

$$\tilde{t} = \Delta\Phi_{i+1,\text{cal}}^{-1}(\Delta\Phi_i) \tag{4.63}$$

$$\tilde{Q} = \Delta\Phi_{i+1,\text{cyc}}^{-1}(\Delta\Phi_i) \tag{4.64}$$

These formulations allow combining arbitrary operating regimes with each other, given that aging trajectories as those shown in Figure 4.5a are available.

4.5.4 End of Life Estimation

Of primary interest with regard to optimal scheduling are the operating costs given in $ per kWh or $ per hour resulting from maintaining a particular cycling regime over a certain period of time. This is equivalent to the question how long the battery can be operated and how much energy can be cycled under a particular cycling regime until reaching its *end of life* (EOL). This information can be used to compute a cost per unit of time or per unit of cycled energy which can be used in the utility function of the scheduling problem. The EOL condition is given by a certain physical quantity reaching a particular value at which the battery is considered to be not usable anymore. For instance, in the case of PEVs, the EOL is commonly assumed to be reached once the battery capacity has decreased to 80 % of its initial value. Another factor determining EOL in the context of PEVs can be considered the power capacity since after a certain power fade the battery may not be able to provide the desired acceleration behavior anymore. Different EOL conditions may hold for stationary batteries which may become unusable once energy losses due to internal resistance increase render their operation unprofitable.

In general terms, EOL can be defined by the function $\Phi(t)$ as defined in Section 4.5.3 reaching a value Φ_{EOL}. The time till EOL can therefore be calculated by

$$\Phi(t_{\mathrm{EOL}}) - \Phi_{\mathrm{EOL}} = 0 \qquad (4.65)$$

If calendar and cycle aging are considered simultaneously, this can be a somewhat complicated equation which has to be solved numerically, e.g., using Brent's method [194].

In case the EOL can be determined by multiple physical properties Φ^1, Φ^2 etc. such as capacity and internal resistance, the actual t_{EOL} is determined by the condition which is reached first so that

$$t_{\mathrm{EOL}} = \min\{t_{\mathrm{EOL},\Phi^1}, t_{\mathrm{EOL},\Phi^2}, ...\} \qquad (4.66)$$

4.5.5 Computing State of Health

In order to assess the value loss of a battery, the physical damage caused to a battery during operation needs to be quantified. This is done by the concept of the SOH. While there is no strict definition of SOH, a reasonable presumption is that it should exhibit the property that it assumes a value of 1 when the battery is new and decreases to 0 as the battery reaches its EOL. A general definition for the SOH χ can therefore be formulated as

$$\chi(t) = \frac{y(t) - y_{\text{EOL}}}{y_0 - y_{\text{EOL}}} \qquad (4.67)$$

which is the relation between the remaining usable value of a physical quantity y at time t compared to its initial usable value at the BOL. In this relation, y may represent a parameter such as capacity, resistance, number of cycles, time or cyclable energy [195].

The problem of many physical quantities such as capacity, however, is that they may not decrease linearly despite constant operating conditions or that they may not even be monotonically decreasing. This causes difficulties when assessing the battery value based on SOH because the value may, for instance, remain almost constant over an extended period of time followed by a sudden drop or may even temporarily increase even though the cell is aging.

For this application, it is desirable that the SOH decreases linearly as long as the operating conditions are constant. As calendar and cycle aging shall be addressed simultaneously, it also needs to contain parameters which depend on both time and energy throughput. This is achieved by using Equation (4.65) which implicitly defines the remaining time till EOL depending on both calendar and cycle aging processes. After computing t_{EOL}, using these equations and applying Equation (4.66), the SOH can be written as

$$\chi(t) = 1 - \frac{t}{t_{\text{EOL}}} \qquad (4.68)$$

This implies an SOH of 1 if the battery is new and an SOH of 0 once the battery has reached its EOL.

4.5.6 Degradation Monetization

To economically efficiently schedule a BESS, knowledge about the marginal costs under any operating conditions at the present state is necessary. This requires monetizing battery depreciation by establishing a relationship between the change of the relevant physical properties with a monetary unit.

In general terms, marginal degradation costs can be expressed by the change of the battery's remaining value, further referred to as the *salvage value* S. In a similar fashion to a consideration in [195], the salvage value for the applications considered in this work can be defined as

$$S(t) = C_{\mathrm{BP}} K(t) \chi(t) \tag{4.69}$$

In this relation, C_{BP} states the purchasing costs of the battery pack and K denotes a discount factor. The discount factor accounts for the value loss which arises from the fact that a used item can typically be sold only at a lower price than the new product.

The marginal value loss of the battery can be described as the change of $S(t)$ according to

$$c' = \frac{\mathrm{d}S(t)}{\mathrm{d}t} \tag{4.70}$$

$$= C_{\mathrm{BP}} \left(\frac{\partial K(t)}{\partial t} \chi(t) + K(t) \frac{\partial \chi(t)}{\partial t} \right) \tag{4.71}$$

The first term in Equation (4.71) addresses the value change resulting from a change in the discount factor which is not of primary interest in this context. The second term is the cost resulting from decreasing SOH which represents physical battery degradation.

The decision on how to operate the battery at a particular time t depends on the marginal cost of cycling. These exclude costs which would incur in any case such as the first term in Equation (4.71) and the cost incurring through calendar aging while idling the battery at the current SOC. The cost driving the cycling decision can therefore be described as

$$c = C_{\mathrm{BP}} K(t) \left(\frac{\partial \chi(t)_{\mathrm{cycling}}}{\partial t} - \frac{\partial \chi(t)_{\mathrm{idling}}}{\partial t} \right) \tag{4.72}$$

With Equation (4.68), this resolves to

$$c = C_{\mathrm{BP}} K(t) \left(\frac{1}{t_{\mathrm{EOL,cycling}}} - \frac{1}{t_{\mathrm{EOL,idling}}} \right) \tag{4.73}$$

In principle, this cost may also become negative as in some cases idling the battery at a high SOC may exceed the cycle costs of discharging.

4.5.7 Degradation Model Parametrization

Up to this point, the SOH is defined generically. For the physical modeling of the SOH, a degradation model presented in [190] is adopted to model the actual capacity decrease and internal resistance increase. In [190], it was found that the calendar capacity loss and internal resistance increase can be described by a $t^{0.75}$ dependency so that

$$\Gamma_{\mathrm{cal}}(t, T, U) = 1 - \alpha_{\mathrm{cap}}(T, U) t^{0.75} \tag{4.74}$$
$$R_{\mathrm{cal}}(t, T, U) = 1 + \alpha_{\mathrm{res}}(T, U) t^{0.75} \tag{4.75}$$

with the temperature T, the voltage U at which the cell is stored and the coefficients α_{cap} and α_{res}. Likewise it was found that the cycle capacity loss as a function of the accumulated charge throughput Q can be described with a square root dependency on Q while the internal resistance increase exhibits a linear dependency so that

$$\Gamma_{\mathrm{cyc}}(Q, \Delta x, \overline{U}) = 1 - \beta_{\mathrm{cap}}(\Delta x, \overline{U}) \sqrt{Q} \tag{4.76}$$
$$R_{\mathrm{cyc}}(Q, \Delta x, \overline{U}) = 1 + \beta_{\mathrm{res}}(\Delta x, \overline{U}) Q \tag{4.77}$$

with the average voltage of a cycle $\overline{U}$ and coefficients β_{cap} and β_{res}. Using these expressions, the capacity and resistance as a function of time and charge throughput for simultaneous calendar and cycle aging are written as

$$\Gamma(t, Q, T, \Delta x, \overline{U}) = 1 - \alpha_{\text{cap}}(T, \overline{U})t^{0.75} - \beta_{\text{cap}}(\Delta x, \overline{U})\sqrt{Q} \quad (4.78)$$

$$R(t, Q, T, \Delta x, \overline{U}) = 1 + \alpha_{\text{res}}(T, \overline{U})t^{0.75} + \beta_{\text{res}}(\Delta x, \overline{U})Q \quad (4.79)$$

The coefficients α_{cap}, α_{res}, β_{cap} and β_{res} were determined by the authors to be

$$\alpha_{\text{cap}}(T, U) = (7.543U - 23.75) \cdot 10^6 \exp(-6976T^{-1}) \quad (4.80)$$

$$\alpha_{\text{res}}(T, U) = (5.27U - 16.32) \cdot 10^5 \exp(-5986T^{-1}) \quad (4.81)$$

$$\beta_{\text{cap}}(\overline{U}, \Delta x) = 7.348 \cdot 10^{-3}(\overline{U} - 3.667)^2 + 7.6 \cdot 10^{-4}$$
$$+ 4.081 \cdot 10^{-3}\Delta x \quad (4.82)$$

$$\Delta\beta_{\text{res}}(\overline{U}, \Delta x) = 2.153 \cdot 10^{-4}(\overline{U} - 3.725)^2 - 1.521 \cdot 10^{-5}$$
$$+ 2.798 \cdot 10^{-4}\Delta x \quad (4.83)$$

Since calendar and cycle aging take place simultaneously, the voltage U as given in Equations (4.74) and (4.80) is replaced by the average voltage $\overline{U}$ in Equations (4.78) and (4.79).

4.6 Conclusion

This chapter formalizes the scheduling problem to be addressed by the framework described in Chapter 3 and introduces the relevant models. The problem under consideration is a scheduling problem where the charging/discharging of a BESS shall be controlled in a way maximizing financial profits. Due to the non-linearity of battery degradation, this problem cannot be solved by a simple linear programming approach. As an alternative to attempting to piecewisely linearize the problem or to a formulation as an MINLP, the problem is addressed using a DP formalism. This comes at the advantage of greater universality also with regard to accounting for the potentially stochastic nature of the problem. It further provides increased performance as compared to an MINLP formulation which is NP hard. As a rolling horizon procedure, this methodology easily handles the non-stationary nature of the problem as updated price forecasting information becomes available or constraints change over time.

The BESS is modeled at cell level using an EC model which relates the electrical behavior of the cell to the power at the terminal of the BESS, the latter of which is the number of interest as seen from a power grid perspective. The EC model considers the OCV characteristic of the cell, the IR drop and the polarization voltage. Polarization voltage is determined based on the difference between measured voltage on one side and OCV and ohmic losses on the other one. It is acknowledged that this phenomenological representation does not attempt to model underlying physical or chemical processes and therefore cannot be considered an attempt to appropriately describe the behavior of a Li-ion battery cell. This would require more thorough investigations including *electrochemical impedance spectroscopy* (EIS) measurements and variations of charging conditions and relaxation times to account for the dynamics of the cell behavior. Since the polarization voltage only makes a very small contribution to the overall voltage of approx. 0.1–4 %, its influence in the context of the presented charge scheduling mechanism, however, is negligibly small. As all experimental results are based on measurements below 1 C, the findings can also only be considered applicable to cases of moderate C-rates. An aspect not considered by the battery model is self-discharge. While self-discharge can play a notable role in various battery types including Ni-MH and Pb-Ac batteries, self-discharge rates in the context of Li-ion batteries are at about 2 % per month [54]. Due to the focus on this type of batteries, this simplification is justified in the presented case. For extensions to other battery chemistries, this aspect may, however, have to be considered.

The degradation monetization model allows computing the degradation costs as a function of the operating parameters and is able to track the cell's SOH in terms of its capacity or internal resistance. An aspect neglected by this representation is the C-rate dependency of degradation. As ohmic losses are proportional to the current, high C-rates typically result in higher temperatures which in turn accelerate aging processes. In the given context, however, typical C-rates can be expected to be very low, mostly far below 1 C. This is a consequence of the current price levels for DSM which do not justify large SOC swings in a short period of time, as indicated by the results presented in Chapter 5. In the range below 1 C, the C-rate dependency of battery degradation is small compared to the impact of other parameters such as the SOC level as argued in [196]. Under these conditions, neglecting the C-rate as an influencing parameter is therefore justified. In case the model is to be applied to scenarios with higher charging powers, this aspect may have to be included as further discussed in Section 6.2.

The degradation model allows tracking the battery's SOH over time, meaning that at any point in time the calendar and cycle contributions to overall degradation are known. This is necessary to compute operating costs and to identify the time when the battery's EOL is reached. For the case that the presented approach would be used to schedule the charging/discharging of a real battery it needs to be noted that the modeled cell characteristics and the real cell's SOH would diverge over time since modeling inaccuracies accumulate as time proceeds. While this is not a concern regarding the studies conducted in this work, this would require an approach for continuously re-calibrating the model in a real-world system.

5 Applications

5.1 Introduction

The case studies presented in this section serve as a demonstration of the developed framework and methodology and shall provide an insight into the financial aspects of mobile and stationary BESSs. Section 5.2 describes the standard parameter values which are common for all simulation studies. In Section 5.3, the primary focus is set on the battery degradation model in order to establish an understanding of its functioning and to investigate the financial implications of battery degradation. Section 5.4 demonstrates the sensitivity of the scheduling approach with regard to the relevant parameters by employing the entire framework including the DP optimization formalism. Section 5.5 then compares the financial implications of different charging strategies in a Singapore context using results generated with CityMoS Traffic. Finally, Section 5.6 investigates the economic viability of BESSs in different markets. The main outcomes are summarized in Section 5.7.

5.2 General Parameters

The BESS is modeled as a battery pack consisting of Li-ion cells in accordance with the reference cell and test cell introduced in Chapter 4. As the EOL is most commonly defined through the capacity [197], this criterion is also applied for EOL determination in the presented studies. For the considerations in Section 5.3, the degradation model is of primary interest. In this section, the charging/discharging efficiency is therefore set to 1 and the energy throughput for charging and discharging is simply calculated according to $\Delta E = (x_n - x_m)C_0$. In Section 5.4, energy throughput and energy losses are then computed based on the EC model presented in Chapter 4.

© Springer Fachmedien Wiesbaden GmbH, part of Springer Nature 2019
D. Pelzer, *A Modular Framework for Optimizing Grid Integration of Mobile and Stationary Energy Storage in Smart Grids*, https://doi.org/10.1007/978-3-658-27024-7_5

Table 5.1: Standard parameter values.

Parameter	Value
Battery pack capacity [kWh]	20
Battery cell capacity [Ah]	2.15
Inverter efficiency	0.98
Max. charge/discharge rate	1
Battery price [$ per kWh]	200
Battery temperature [K]	303.15
Relative EOL capacity	0.8

Table 5.1 lists the parameter values which, unless stated otherwise, apply to the entire chapter. The battery pack capacity of 20 kWh is a typical value for a range of compact PEVs such as the BMW i3, Nissan Leaf and Mitsubishi i-MiEV. The capacity of the single battery cell of 2.15 Ah corresponds to the reference cell used for the degradation model. In order to compute the number of cells a battery pack with the indicated capacity is composed of, the EC model is used by determining the battery cell's energy content during a full discharge at 0.01 C. Apart from the battery cells, an essential component of a battery pack is an inverter of which the efficiency is set to 0.98 in accordance with [198]. Combined with the charging losses provided by the EC model, this leads to an efficiency of 0.85 for an entire charge-discharge cycle. Setting a maximum charge/discharge rate is somewhat arbitrary because it depends on the specifications of the battery as well as the type of the power connection. The chosen maximum C-rate of 1 C is a value for which the battery model can be considered valid and which in the case of a 20 kWh battery corresponds approximately to an AC Level 2 connection with a maximum charging power of 19.2 kW. The battery temperature of 303.15 K is assumed to be a realistic value considering that temperatures' during operation with thermal management may be somewhat above this value while temperatures during idling may be somewhat lower. The battery price of $ 200 per kWh corresponds to a price target which

is expected to be reached before the end of the decade as discussed in Section 2.4. The EOL capacity of a factor 0.8 compared to the capacity of the new battery pack is in accordance with the definition in [162] and is a common assumption for PEV batteries [197]. As many of these parameters can be subject to variations, sensitivity considerations are performed in this chapter.

5.3 Understanding the Implications of the Degradation Model

In this section, the degradation model presented in Section 4.5 is employed in a number of different scenarios. The purpose of these considerations is to demonstrate the functionality of the model, to create a sense for its sensitivity with regard to the relevant parameters and to provide an insight into the magnitude of the model outputs of interest. This shall also simplify the interpretation of the subsequently discussed results. These investigations cannot cover every possible parameter combination, instead, a number of instructive examples are chosen which are believed to provide a qualitative and quantitative understanding of the implications of battery degradation.

5.3.1 Capacity Loss and the Influence of Calendar and Cycle Degradation on Costs

In this section, the decrease of capacity in relation with the cycling conditions and the change of the salvage value are investigated. Furthermore, the costs as a function of the operating parameters are computed. As this serves for illustration purposes, an artificial set of simple cycling conditions is used. For Figure 5.1, the cell is assumed to be cycled between the two SOC values 0.45 and 0.55, further denoted as $(0.45, 0.55)$, until the capacity reaches 90 % of its initial value. Subsequently, cycling proceeds in a range $(0.25, 0.75)$. The C-rate is kept constant at 1 C. Figure 5.1a shows the remaining capacity as a function of time. The capacity curve starts at a value of

1 and decreases up to a value of 0.8 which, in this case, is considered the EOL condition. The break at 90 % indicates the point where the cycling regime changes from $(0.45, 0.55)$ with an SOC swing of 0.1 to $(0.25, 75)$ with an SOC swing of 0.5. The capacity decline is non-linear since Equations (4.74) and (4.76) are both non-linear in time and charge throughput. This shows the problem which would occur if cycle costs were quantified proportional to capacity decrease since the same amount of cycled energy would incur different costs at different points of the battery's lifetime. As shown in Figure 5.1b, the computation of the salvage value through the time till EOL as proposed in Section 4.5.6 resolves this issue. As requested in Section 4.5.6, the salvage value decreases linearly under constant cycling conditions which implies constant costs per unit of cycled energy.

The slope of the salvage value decrease changes at the point where the cycling conditions change. This is because the new cycling conditions with a greater SOC swing cause accelerated aging as follows from the Δx dependency of Equation (4.82). Under these conditions, the remaining value of the battery pack has to be distributed over a smaller amount of cycled energy than under the first set of cycling conditions. This increases the costs per cycled unit of energy. Each cycling regime is therefore priced according to how much degradation it imposes on the battery. As shown in Figure 5.1c, even though the battery loses 10 percentage points of capacity in each cycling regime, 63 % of the energy throughput occur in the first one. At the same time, the battery pack loses only 28 % of its value in the first cycling regime while 72 % are lost in the second one. With $0.02 per kWh versus $0.086 per kWh, degradation costs per unit of cycled energy are therefore 4.3 times higher in the second cycling regime. The discrepancy would increase further with a smaller Δx in the first cycling regime and a larger Δx in the second one. This already indicates the problems arising from assuming an average degradation cost without precise knowledge on the actual cycling conditions.

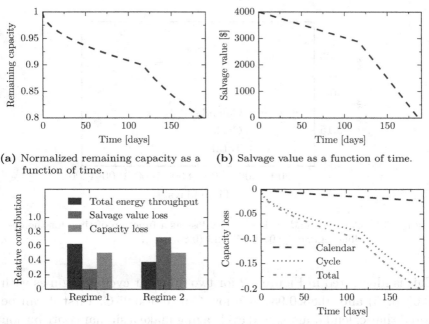

(a) Normalized remaining capacity as a function of time.

(b) Salvage value as a function of time.

(c) Comparison of total energy throughput, decrease of salvage value and capacity loss.

(d) Calendar and cycle capacity loss as a function of time.

Figure 5.1: Various model output parameters for a sequence of two different cycling regimes $(0.45, 0.55)$ and $(0.25, 0.75)$.

Figure 5.1d shows the capacity loss over time for calendar and cycle aging separately. As calendar and cycle degradation are treated as separate processes, both curves are continuous also when operating conditions change. In this case, the major contribution to degradation stems from cycle aging with calendar aging only contributing 11.2 % to overall capacity loss. Calendar aging capacity loss remains fairly linear because $\overline{U}$ in Equation (4.80) remains unaffected by the change of operating conditions and the non-linearity of Equation (4.74) in t with an exponent of 0.75 is weak. In this case, calendar degradation is therefore relatively unimportant.Even though calendar degradation plays a small role in the presented case this is not generally true.

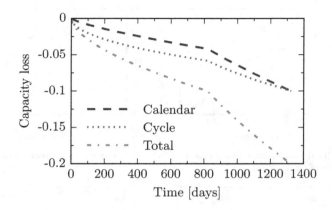

Figure 5.2: Calendar and cycle capacity loss as a function of time for the cycling regimes $(0.05, 0.15)$ and $(0.85, 0.95)$.

This is illustrated in Figure 5.2 for two different cycling regimes with $(0.05, 0.15)$ and $(0.85, 0.95)$ at a low C-rate of $0.05\,C$. First it can be noted that calendar aging and cycle aging make a similar contribution to overall capacity loss. The simple reason for this is that due to the lower C-rate a smaller amount of energy is cycled which slows down cycle aging compared to the first scenario. The calendar degradation processes therefore have more time to manifest themselves. It is also observed that calendar aging proceeds faster for the second cycling regime in the SOC range $(0.85, 0.95)$. The reason for this is the voltage dependency of Equation (4.80) and the fact that a higher SOC comes with a higher voltage. In this case, neglecting calendar degradation would lead to a considerable error with regard to the determination of the battery's EOL.

As shown in Figure 5.3, the scenario with the low C-rate of $0.05\,C$ (Scenario 2) implies a notably higher lifetime of 3.6 years compared to 0.5 years for the previously considered scenario with a C-rate of $1\,C$ (Scenario 1). This is the consequence of the lower energy throughput which only amounts to $36\,\%$ compared to Scenario 1. As a smaller amount of energy has to be distributed over the same overall battery

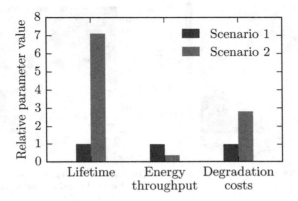

Figure 5.3: Comparison of various model outputs for two investigated scenarios. Scenario 1 consists of the cycling regime $(0.45, 0.55)$ followed by $(0.25, 0.75)$ and Scenario 2 of the regime $(0.05, 0.15)$ followed by $(0.85, 0.95)$.

cost, the average cost of cycled energy is 2.8 times higher in Scenario 2, corresponding to a value of \$ 0.13 per kWh.

Since the results in Section 5.3.1 only represent an example for a particular operating regime, Figure 5.4 draws a more complete image of degradation costs and contribution of cycle and calendar aging as a function of the cycling regime. The values are computed for an SOC discretization of 0.02 and x_1 and x_2 indicate the start and end SOC of the respective cycling regime. Figure 5.4a shows the degradation costs resulting from cycle aging only. It can be seen that costs generally increase with SOC swing due to the Δx dependency of Equation (4.82). Costs are minimal at moderate SOC levels with small SOC swings leading to costs smaller than \$ 0.01 per kWh and maximal at extreme SOC values with large SOC swings leading to costs up to \$ 0.21 per kWh. Due to the polynomial voltage dependency of Equation (4.82), costs are somewhat higher at SOCs at the upper end of the scale than at the lower one.

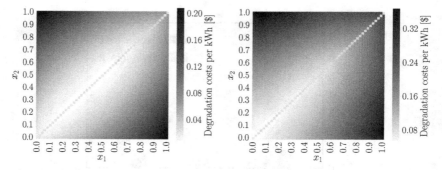

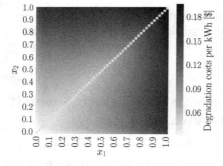
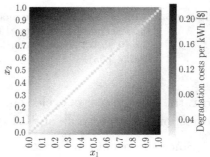

(a) Degradation costs per kWh for cycle aging only.

(b) Combined degradation costs per kWh for cycle and calendar aging at C-rate of 0.05 C.

(c) Difference between degradation costs without and with considering calendar degradation at C-rate of 0.05 C.

(d) Combined degradation costs per kWh for cycle and calendar aging at C-rate of 1 C.

Figure 5.4: Degradation costs and contribution of aging processes to overall degradation for different state transitions.

Figure 5.4b shows degradation costs when considering both calendar and cycle aging at a C-rate of 0.05 C. The difference between the values in Figure 5.4b and Figure 5.4a is shown in Figure 5.4c. In Figure 5.4b, it can be seen that high SOC states become considerably more expensive than low SOC states which results from the voltage dependency of Equation (4.80). In this case, average costs increase from \$ 0.067 per kWh to \$ 0.167 per kWh. The maximum cost which is the cost of a full charge or discharge increases from \$ 0.21 per kWh to

$0.35 per kWh. The reason for this considerable cost increase is not simply the addition of calendar aging costs which for an idle battery at the given temperature would only average $0.05 per hour of storage. Instead, as becomes clear from revisiting Figure 5.1d, the reason is that the battery price has to be divided by a smaller lifetime energy throughput than without calendar aging. In case calendar aging is not considered, the time till EOL is computed through the dotted curve in Figure 5.1d. Since this curve reaches the EOL condition much later than the correct dash-dot line, the time till EOL and therefore the possible amount of energy which can be cycled are largely overestimated. This leads to an underestimation of degradation costs per kWh. The major effect of calendar aging on the cost per kWh therefore is not the additive cost of calendar aging. It is, in contrast, the shortening of the battery lifetime which limits the lifetime energy throughput and therefore requires distributing the battery price over a smaller amount of energy. The shorter the calendar lifetime, the higher the cycle costs per kWh since the same battery price is distributed over a smaller amount of energy.

The observed increase in costs due to calendar degradation is, of course, notably dependent on the C-rate. If the cell is continuously subject to heavy cycling, the greatest contribution to capacity loss stems from cycle aging since the cell does not have enough time for calendar aging. This can be seen in Figure 5.4d for a C-rate of 1 C. In this case, the difference to Figure 5.4a is negligibly small.

As Equation (4.80) is temperature dependent, the contribution of calendar aging depends on temperature as well. The dependency of calendar degradation costs only on temperature for different SOCs is shown in Figure 5.5. For low temperatures and low SOCs, costs are negligibly small whereas high temperatures and high SOCs result in costs up to $0.7 per hour. Table 5.2 lists costs per kWh averaged over all SOC transitions in the case of calendar and cycle aging for different temperatures and different C-rates. While temperature has a considerable influence in the case of small C-rates where

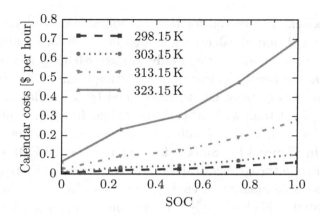

Figure 5.5: Calendar costs per hour for 20 kWh of battery capacity at different temperatures.

calendar aging plays a dominant role, the effect is rather moderate in the case of a high C-rate. Nevertheless, this also indicates a certain relevance of appropriate measures to control temperature during operation.

Table 5.2: Average cost per kWh and the corresponding standard deviation in parentheses for different temperatures and C-rates.

| | Avg. cost and std. dev. [$ per kWh] | |
Temperature [K]	0.05 C	1 C
– (cycle aging only)	0.064 (0.046)	0.064 (0.046)
298.15	0.136 (0.065)	0.071 (0.048)
303.15	0.173 (0.076)	0.075 (0.049)
313.15	0.303 (0.115)	0.087 (0.052)
323.15	0.579 (0.202)	0.110 (0.058)

5.3.2 Dependency of Degradation Costs on Other Battery Parameters

There are a number of additional parameters which may have an effect on degradation costs. These include the battery price, the battery capacity and the EOL criterion. In this section, the sensitivity of degradation costs with regard to these aspects is illustrated. As these relationships are fairly straightforward, they are all summarized in Figure 5.6. Battery prices are varied between $100 per kWh and $500 per kWh, capacity between 20 kWh and 100 kWh and $\Delta\Gamma_{EOL}$ between -0.2 and -1.0, all in five equidistant steps. In order to present all three parameters in one diagram, the x-labels are normalized so that the smallest parameter value (the largest in the case of $\Delta\Gamma_{EOL}$ because these values are negative) is given by 1. Degradation costs increase linearly with slope 1 for variation of battery prices which is simply rooted in the fact that the battery costs are divided by the lifetime energy throughput to compute the cost per unit of energy. The capacity is irrelevant for degradation costs because increasing the capacity increases battery investment costs and energy throughput by the same factor. The only interesting parameter in this context is the EOL criterion which has a non-linear effect on degradation costs. As shown in Figure 5.7, it is linear in a log-log plot so that it follows a relation $f(x) = c\Delta\Gamma_{EOL}^m$. The reason for this decrease is that the lifetime energy which can be cycled through the cell increases when the cell is used for a longer time. While increasing the operation time only would result in a linear increase of the energy throughput, the flattening of the capacity loss according to the parametrization provided in Section 5.2 leads to a superlinear cost decrease. It has to be noted, however, that the deployed model does not describe the sudden death of the cell which eventually leads to a steep capacity decrease once a certain state of degradation is reached. Beyond the validity domain of the presented model, the decrease in degradation costs would therefore be less pronounced.

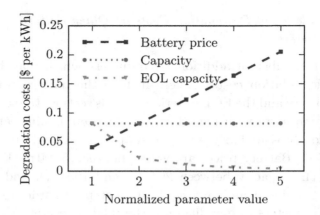

Figure 5.6: Average degradation costs per kWh considering calendar and cycle aging by varying battery price, battery capacity and the EOL condition.

Another parameter which so far has not been considered is the efficiency for charging and discharging which was set to 1 for all previous considerations. The efficiency can be computed on a low level based on the EC model presented in Section 4.3. For the consideration

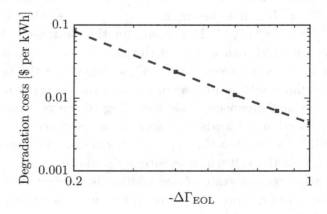

Figure 5.7: $\Delta\Gamma_{EOL}$ dependency of calendar and cycle costs in a log-log plot.

presented here, however, a simple approximation is sufficient: the efficiency of the cell is determined by the coulombic efficiency as well as the voltaic efficiency. The former describes the charge losses occurring during chemical energy conversion. This efficiency is typically around 0.999 [54] and therefore negligible. Voltaic efficiency refers to thermal losses occurring due to ohmic losses as well as charge transfer and diffusion resistance. These losses typically result in a cell efficiency in the range of 0.95–0.98 [54] at around 1 C. As indicated in [199], the dependency of efficiency on SOC is weak for Li-ion batteries which is why efficiency is assumed to be constant over SOC. On a battery pack level, additional inverter losses need to be taken into account. In the considered power range, the power converter efficiency can be considered to be constant at a value of approx. 0.96–0.98 for both current directions [198, 200, 201]. This results in a lower boundary of 0.91 and an upper boundary of 0.96 of overall efficiency. With an electricity price of \$ 0.15 per kWh which was the residential *electricity tariff* (ET) in Singapore in the third quarter of 2017 [92], this results in losses of \$ $6 \cdot 10^{-3}$ per kWh to \$ 0.014 per kWh. An example with relatively high electricity prices is Germany where the average household electricity price of the year 2016 was \$ 0.32 per kWh [202]. This leads to losses of \$ 0.013 per kWh to \$ 0.029 per kWh. Comparing these values with Table 5.2 shows that the energy losses at the upper end of this range are still at the lower end of the spectrum of degradation costs. In most cases, battery degradation costs are therefore the dominant cost factor.

5.3.3 Conclusion

The presented considerations provide a first insight into the effect different operating conditions can have with regard to the relevance of the different degradation processes. A main conclusion from the presented results is that neither calendar aging nor cycle aging can generally be neglected. Instead, the relevance of the respective process highly depends on the specific operating conditions. While calendar aging plays a comparably small role in case the battery is subject to heavy cycling, the contrary is true if the battery is kept idle for extended periods of time or if it is consistently operated at low C-rates. When deploying BESSs for grid applications, C-rates can generally not be expected to be extraordinarily high. Especially in these cases which are of primary interest in this work, calendar degradation can therefore be assumed to play a relevant role. The same applies to PEVs which are parked approx. 95 % of the day [49]. Also in this case, charge optimization with regard to calendar degradation could therefore be expected to result in notable savings. This hypothesis is further investigated and confirmed in Section 5.5.

The high variability of degradation costs which can spread across two orders of magnitude for different operating conditions also shows that the assumption of constant degradation costs taken by many studies listed in Chapter 2 is problematic. Justifying this assumption requires precise knowledge on the average cycling conditions in order to keep the error of averaging within an acceptable range. Cycling conditions, however, highly depend on a number of framework conditions including electricity prices and battery degradation behavior itself. Reliable estimates in this regard are therefore hard to make.

The presented considerations thus show that not appropriately accounting for degradation costs when attempting to compute optimal charging/discharging schedules is likely to yield undesired results. As a result which is in accordance with the related work cited in Section 2.6.2, degradation costs can therefore by far outweigh possible revenues attainable in electricity markets if battery aging is not appropriately considered. At the same time, however, the results show that due to the high variability of degradation costs depending on the operating conditions, a degradation-aware scheduling mechanism could considerably reduce degradation costs. This may open a window of opportunity for operating a BESS in a profitable way. These observations therefore underline one of the main assumptions underlying this work that appropriately optimizing battery scheduling with regard to degradation costs is a crucial factor for profitable operation.

5.4 Parameter Sensitivity

There is a variety of parameters exerting influence on the optimization outcome. In order to assess results obtained with the presented method with regard to these parameter sensitivities, this section investigates an energy arbitrage scenario with regard to the profits which can be attained under a variety of conditions and parameter settings. While Section 5.3 investigated the degradation model in isolation, the results in this section are based on deploying the entire framework including the DP scheduling algorithm and the EC model presented in Chapter 4.

5.4.1 General Parameters

The optimization lookahead of the DP solver is set to a value of 7 and the state discretization to 0.05. As shown in this section, this is a reasonable compromise between accuracy and computing time. Unless noted otherwise, the general parameter values listed in Section 5.2 apply. As example data, 1 500 periods of generation prices from the *National Electricity Market of Singapore* (NEMS) are used for both buying and selling energy[1]. In this data set, the top 1 % of the prices are set to the price average since these outlier periods would otherwise dominate the entire outcome.

5.4.2 Results

This section shows the sensitivity of the optimization outcome with regard to the way of quantifying degradation costs, lookahead, state

---

[1] The data covers the periods 5 000-6 499 of the year 2009. This data is chosen because price levels and price variability in these periods are relatively high so that a sufficient amount of cycling is ensured.

space discretization, battery price, EOL condition, temperature and lower SOC threshold.

5.4.2.1 Effects of Different Degradation Cost Quantification Methods

This investigation explores how different ways of quantifying battery degradation costs affect profitability. The scenarios include degradation cost consideration using i) the degradation model established in this work with both calendar and cycle aging, ii) the same model without considering calendar aging, iii) no consideration of degradation at all and iv) different constant degradation costs per kWh. For constant degradation costs, three different values of \$ 0.012 per kWh, \$ 0.075 per kWh and \$ 0.173 per kWh are chosen. The first one is an optimistic value as obtained from the data underlying Figure 5.4d. The second and third value correspond to the average cost for cycling at 1 C and 0.05 C corresponding to the results shown in Figure 5.4d and Figure 5.4b.

The results of all optimizations are listed in Table 5.3. The first data column shows the simulated profit as it is computed when conducting the optimization. The second data column results from taking the simulated charge/discharge cycles as an input for computing the 'real' degradation costs using the full degradation model as described in Chapter 4. It therefore quantifies the real profit which would have been made for the scenarios i) through iv). The third and fourth data columns inform about the corresponding simulated and real value loss of the battery pack. The value loss describes the monetary depreciation of the battery due to calendar and cycle degradation during the investigated time horizon.

The first two rows in Table 5.3 show the result for the standard degradation model with and without considering calendar aging. The simulated profit is higher in the case without calendar aging. At the same time, the simulated value loss without calendar aging is lower. The real value loss of the battery pack, however, exceeds the simu-

Table 5.3: Simulated and real profits resulting from different ways of considering battery degradation costs.

Model parametrization	Profit [$]		Value loss [$]	
	Sim.	Real	Sim.	Real
Standard degrad. model				
Cal. and cyc. degrad.	81	81	-71	-71
Cyc. degrad.	106.9	80.4	-50	-76.4
No degrad.	246.2	-292.3	0	-538.5
Constant cost				
$ 0.012 per kWh	195.1	-104.8	-35.9	-335.7
$ 0.075 per kWh	82.5	33.6	-79.7	-128.6
$ 0.173 per kWh	24.1	35.4	-50.9	-39.6

lated value loss in case no calendar degradation is considered. This leads to a smaller real profit when calendar degradation is neglected. The difference between real profits with and without accounting for calendar aging in this case is small because in the considered scenario the battery automatically maintains a low SOC level, thus keeping calendar degradation low. This partly optimizes the scheduling also with regard to calendar aging 'by accident'. Nevertheless, in the long run the simulated value loss without considering the contribution of calendar aging would be increasingly far off the real value so that the optimization would be based on a wrong cost calculation.

The situation is more extreme if no degradation is considered since in this case in the short period of time under consideration about 14 % of the battery's value are destroyed. This results in a real loss of $ 292.3 while appropriately considering battery wear would result in a profit of $ 81.

Within the constant cost scenarios, profits and value losses vary throughout the different scenarios. Lower degradation costs imply that more cycling is being done with higher revenues resulting in higher simulated profits. At the same time, low degradation costs result in a low simulated value loss. By looking at the second data column, however, it is revealed that the real value losses are far off the simulated values. In the low cost scenario where most cycling is performed, the battery loses $ 335.7 of its value resulting in a total loss of $ 104.8. Also in the medium cost scenario, degradation costs are underestimated so that the battery is cycled too heavily. In the high cost scenario, cycling happens too conservatively so that the battery is treated too carefully. This results in the fact that a relevant part of the possible revenues remain untouched so that the real profit only makes up 41 % of the optimal profit. These results show that an optimization based on average cost considerations is very unlikely to yield a close-to-optimal outcome.

5.4.2.2 Lookahead

The lookahead determines how much forecasting data is available to the optimization algorithm. It is generally limited by the horizon for which sufficiently accurate price data can be obtained under realistic conditions. As runtime increases with the lookahead, a trade-off between optimality and computational cost also needs to be considered. It is therefore important to know whether existing forecasting horizons are sufficient for the method to effectively leverage on price forecasts. Figure 5.8a shows the effect the lookahead has on overall profits. In the region below approx. 5 periods, the graph exhibits a considerable slope implying that in this domain additional information has a significant marginal benefit. Starting from approx. 6–7 periods, profits saturate so that additional information does not lead to a better optimization outcome. The lookahead is therefore set to 7 periods in the remainder of this work, unless stated otherwise.

5.4.2.3 State Space Resolution

The resolution of the state space discretization determines how fine the optimizer can tune the control variable. At the same time, runtime increases with the number of states. Figure 5.8b shows how profits depend on the state space resolution. It can be seen that there

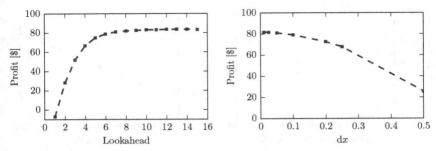

(a) Profit as a function of the lookahead.   **(b)** Profit as a function of the state discretization.

Figure 5.8: Sensitivity with regard to solver settings.

is little variation in the range below $dx = 0.1$ which corresponds to 11 possible states. This is because very small SOC swings are made only when prices are low. As low prices with small amounts of cycled energy also imply low revenues, the periods with small SOC swings have a low weight with regard to overall profits. With discretizations coarser than 0.1, profit decrease becomes steeper, meaning that an increasing fraction of the optimization potential remains unexploited. For a reasonable compromise between runtime and optimality, discretization is set to 0.05 in the remainder of this work, unless stated otherwise.

5.4.2.4 Battery Price

As discussed in Section 2.4, battery prices have been dropping considerably in the past and a further decrease is expected. In Section 5.3, it was shown that degradation costs per kWh scale linearly with battery prices. Cutting battery prices by half therefore also cuts degradation costs by half. As shown in the logarithmic plot in Figure 5.9a, profits decrease exponentially with increasing battery prices. Qualitatively speaking, this is because reducing the battery price proportionally reduces degradation costs per unit of energy cycled which, in turn, results in a greater overall energy throughput. As a consequence, fu-

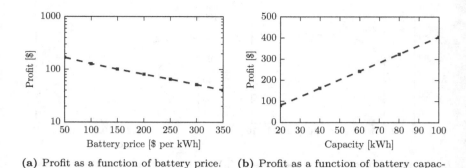

(a) Profit as a function of battery price.   **(b)** Profit as a function of battery capacity.

Figure 5.9: Sensitivity with regard to fundamental battery specifications.

ture drops in battery prices can exhibit a comparably large effect on attainable profits.

5.4.2.5 Capacity

As already discussed in Section 5.3, the capacity does not have any effect on the cost parameters. As the total energy throughput scales linearly with the capacity, profits also linearly depend on capacity as shown in Figure 5.9b.

5.4.2.6 End of Life Condition

Due to the non-linear decrease of degradation costs when relaxing the EOL condition, profit considerably increases the longer the battery is used. This is illustrated in Figure 5.10a. Relaxing $\Delta\Gamma_{EOL}$ from -0.2 to -0.4 therefore has the same effect on profitability as decreasing the battery price from $\$\,200\,per\,kWh$ to $\$\,50\,per\,kWh$. This shows that relaxed EOL conditions, which are realistic even for PEVs according to [197], can have a notable impact on profitable operation. It has to be noted that the plotted results are only valid to the point where the cell experiences a sudden death so the low values of $\Delta\Gamma_{EOL}$ need to be handled with care.

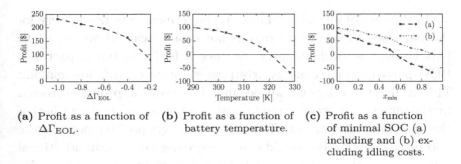

(a) Profit as a function of $\Delta\Gamma_{\text{EOL}}$.

(b) Profit as a function of battery temperature.

(c) Profit as a function of minimal SOC (a) including and (b) excluding idling costs.

Figure 5.10: Profits as a function of various battery operation parameters.

5.4.2.7 Temperature

As shown in Figure 5.10b, profits near the lower bound of the investigated range are fairly insensitive with regard to temperature. This is because of the temperature dependency of Equation (4.80) which results in slow calendar aging at low temperatures. However, starting from a temperature above 320 K the cell degrades so fast that revenues are not sufficient to compensate for calendar capacity loss.

5.4.2.8 Minimal State of Charge

The minimal accessible SOC x_{min} limits the range which is available for cycling. In reality, this can be defined by a PEV owner who does not want the SOC to drop below a certain threshold. As shown in the dashed line in Figure 5.10c, profits decrease with increasing x_{min}. This has to be expected for two reasons. The first reason is that revenues decrease because an increasing threshold leaves less freedom for cycling and therefore cuts off revenues where large cycle depths would be desirable. Secondly, the system is pushed towards higher SOCs where calendar aging is accelerated as shown in Section 5.3. This leads to the situation that beginning from an SOC threshold of approx. 0.6, energy arbitrage revenues are overcompensated by calendar degradation costs so that profits turn negative.

In the case of negative profits, the investment into the battery for the purpose of energy arbitrage is not economical. However, if the battery pack is available anyways since it is, for instance, part of a PEV, participating in energy arbitrage is not necessarily loss-making. This is illustrated by the dotted line. It shows the profits corrected by the idling costs which would occur in any case, no matter whether the battery is used or not. This value is always positive because the optimizer keeps the battery idle in case cycling would incur additional losses. Using the battery for power grid services can therefore also be economical if the calculated profit value is negative in case the battery investment was made for a different primary purpose. In this case, the primary question is whether effective profits are high enough to justify transaction costs as well as possibly required additional technical equipment for allowing bi-directional energy flows.

5.4.3 Conclusion

Investigating the degradation model in the context of the scheduling mechanism confirms the assumption that computing schedules based on average degradation prices leads to far sub-optimal results, including considerable losses. From this investigation it can be concluded that an appropriate degradation model as part of a scheduling mechanism is a mandatory requirement for profit maximization. The results presented here also show that under the investigated conditions representing a high-price period, operation can be profitable.

The results also show that near-optimal results can be achieved with moderate lookaheads of about 7 periods. For this forecasting horizon, sufficiently accurate information can be obtained in a great number of electricity markets. With regard to the other parameters, the profit dependency of battery price and EOL condition can be considered the most relevant ones. Since the further development of battery systems will specifically affect prices and lifetimes, notable increases in profitability can therefore be expected. This especially makes a case for second life batteries which can come at prices around

$100 per kWh [203] and where EOL conditions are not as strict as in the case of PEVs.

5.5 Comparison of Charging Strategies in a Singapore Context

The leading questions of this section are concerning the financial implications resulting from employing different charging/discharging strategies for PEVs in a Singapore context. Questions addressed in this section include:

- What role do energy costs and battery degradation costs play with regard to the total operating costs of a PEV?

- To what degree can these cost factors be influenced by adopting a smarter charging strategy, i.e., through charging in low price periods and by optimizing the charging schedule with regard to battery degradation?

- How much additional revenue can a PEV owner generate from conducting energy arbitrage through peak shaving and valley filling?

- How do these charging strategies compare to the operating costs of an ICEV?

The study is based on simulations quantifying the charging needs of Singapore drivers based on their mobility behavior. This is done by using CityMoS Traffic as described in Section 3.5.1 to simulate PEVs' driving times and energy consumption for the case of Singapore. This provides information on how much energy each PEV needs to charge in the course of a day and during what times it can be connected to a charging station. Using this information, the economics of different charging strategies can be investigated. In this analysis, the following three charging strategies are compared:

- **Uncoordinated charging**
  In this case, the charging process starts as soon as the PEV is connected to the power grid. The current is set to the maximum

possible value in accordance with the cell's C-rate limitations and with the need for reducing current during CV charging. The charging process proceeds until a certain SOC is reached or the PEV is unplugged from the charging station. Energy is charged at a constant electricity rate as common in retail markets. This means that the charging strategy does not leverage on possible savings from CPP, TOU pricing and real-time pricing as discussed in Section 2.3. Furthermore, the charging process is not optimized with regard to battery degradation. To quantify battery degradation costs for comparison with other charging strategies, these are computed ex post using the models presented in Chapter 4.

- **Smart charging**
  The smart charging strategy employs the methodology presented in Chapter 4, i.e., the charging decision depends on the current and future electricity prices with a certain lookahead. This corresponds to a setting where the PEV is participating in a distributed control DSM scheme implemented through real-time prices as described in Section 2.3. In this case, smart charging only allows uni-directional power flows, thus representing the setting of uni-directional V2G as introduced in Chapter 2. In addition to real-time prices, battery aging is also considered.

- **Energy arbitrage**
  This strategy employs the same method as the smart charging strategy with the difference that bi-directional power flows are permitted. In the terminology of Chapter 2, this is therefore a bi-directional V2G scenario. This method minimizes battery degradation costs, cuts electricity costs through participation in a distributed control setting and generates additional revenues by performing energy arbitrage for the purpose of peak shaving and valley filling.

Table 5.4: Daily trip characteristics simulated with CityMoS Traffic.

Characteristic	Value
Number of trips per agent	2.16
Driving time [min]	28.8
Driving distance [km]	23.1
Energy consumption per 100 km [kWh]	12.8

5.5.1 Data and Parameters

Trip and energy consumption data are generated using the simulation platform CityMoS Traffic. For this purpose, 20 000 agents are simulated based on HITS 2012 data for an entire day by tracking trip times, parking locations and energy consumption. The simulation includes a warm-up period of one day so that the results of the second day are used for the evaluation. Some relevant high-level trip characteristics as simulated by CityMoS Traffic are listed in Table 5.4.

As the optimization framework is designed to handle discrete-time problems, the obtained trip and energy consumption is discretized into periods with a duration of 30 min. This period length is chosen because it corresponds to the time resolution of electricity prices at NEMS. In the uncoordinated charging scenario, two different target SOCs of 0.8 and 0.5 are considered. In the smart charging and arbitrage scenarios, the minimum allowed SOC is set to 0.2 to account for a safety buffer which may be required by the driver.

In this setting, charging spots are assumed to be located at every parking location. As elaborated in further detail in Section 5.5.2, for this study the battery capacity is initially determined based on the daily energy needs computed by CityMoS Traffic in order to match the requirements of a Singapore context. All other parameters not explicitly specified in this section are in accordance with those specified in Section 5.2.

The price data is taken from the Singapore electricity market NEMS. The current market implementation requires non-contestable consumers to pay the fixed ET which is offered in the retail market. This price is used for the uncoordinated charging strategy. As a price-responsive charging strategy is obsolete in the case of a fixed electricity rate, the investigations of the smart charging strategy and the energy arbitrage scenario are based on the *uniform Singapore electricity price* (USEP). The USEP is defined as the weighted average of the nodal wholesale electricity prices which are updated in intervals of 30 min. Since the USEP is the price paid to electricity generators, it does not contain additional cost factors such as taxes and transmission costs. In order to account for these costs, the average difference between ET and USEP is added to every kWh of purchased energy. For the arbitrage scenario, this difference is only added to the energy being consumed and not for the energy which is fed back into the grid. This ensures that taxes and transmission costs are only paid once by the end consumer. For the comparison between PEV and ICEV costs, a gasoline price of $ 1.48 per liter[2] and an average fuel consumption of 7.3 l per 100 km [204] are assumed. In order to keep the number of computations within a reasonable limit, two different weeks, one representing a high-price period and the other one a low price period are simulated. As indicated in Table 5.5 showing averages and standard deviations of the USEP in different years, price levels were high in the year 2013 and low in 2017. Therefore, as a high-price period week 41 of the year 2013 is chosen. With an average price of $ 0.125 per kWh and a standard deviation of $ 0.044 per kWh, this can be considered a fairly typical week of the corresponding year. As the second data set, prices of the week 15 of the year 2017 are taken. This week exhibits a year-typical average price of $ 0.059 per kWh and a standard deviation of $ 0.01 per kWh. For calculating charging costs using the three different strategies, a randomly selected sample of 3 000 agents is used.

---

[2] Corresponds to the average Singapore gasoline price of the first half of 2017 according to https://tradingeconomics.com.

Table 5.5: Average and standard deviation of the USEP in different years. Data for 2017 covers the first 6 months of the year.

Year	USEP [$ per kWh] Avg.	Std. dev.
2013	0.125	0.055
2014	0.099	0.034
2015	0.070	0.062
2016	0.046	0.031
2017	0.059	0.018

5.5.2 Determining Battery Capacity

The battery capacity is determined by employing a simulation-based approach using the framework CityMoS Traffic. The entire setup for this analysis was presented as part of a publication prepared in the context of this work in [8]. For this investigation, the energy consumption of PEVs in Singapore is simulated with the parameters described in Section 5.5.1. The simulation results are then analyzed with regard to the energy consumption of the individual agents. This is illustrated in Figure 5.11 which shows the cumulative share of the PEV population with regard to its daily energy consumption. It can be seen that 95 % of all agents use less than 7 kWh of energy per day with only 1 % requiring more than 10 kWh. Taking this number, 99 % of all daily itineraries could be performed with a 10 kWh battery on a single charge. This is a remarkable result because it refutes the costly trend towards ever larger battery capacities, at least in the context of an urban environment like Singapore and without accounting for behavioral aspects such as range anxiety. The following study therefore uses a capacity of 10 kWh. As it is acknowledged that drivers may require an additional safety buffer to accommodate possible outliers in energy consumption, a second scenario with a battery capacity of 20 kWh is also considered. Furthermore, the implications of greater battery capacities are discussed in order to account for the trend of further increasing battery sizes. For the 10 kWh battery, the

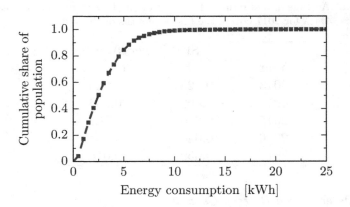

Figure 5.11: Cumulative share of the population according to daily energy consumption as simulated with CityMoS Traffic.

discretization for the solver is set to 0.05 and for the 20 kWh battery it is set to 0.025. These different discretizations are necessary to guarantee that the same mobility pattern results in the same energy consumption for both battery capacities because energy consumption is rounded according to this discretization.

5.5.3 Results

This section presents the results for uncoordinated charging, smart charging and energy arbitrage as well as a comparison between PEV and ICEV costs.

5.5.3.1 Uncoordinated Charging

The aggregated results of the uncoordinated charging strategy are listed in Table 5.6. Results are shown for the years 2013 and 2017 for the two battery capacities of 10 kWh and 20 kWh. Energy costs are simply the amount paid for electricity over the optimization horizon. Idling costs sum up the costs which would incur if the battery was kept idle during each period in the state it assumed at the beginning

Table 5.6: Costs of the uncoordinated charging strategy for one week.

Year	2013				2017			
Batt. cap. [kWh]	10		20		10		20	
Target SOC	0.8	0.5	0.8	0.5	0.8	0.5	0.8	0.5
Energy costs [$]	4.1	3.9	4.0	4.0	3.4	3.2	3.3	3.3
Degrad. costs [$]	8.7	5.2	15.1	8.7	8.7	5.2	15.1	8.7
Idling costs [$]	6.5	3.8	12.9	7.5	6.5	3.8	12.9	7.5
Total costs [$]	12.8	9.1	19.1	12.7	12.1	8.4	18.4	12

of the period. It is therefore the cost resulting from calendar aging only. In contrast, degradation costs are the real costs resulting from battery operation, including calendar and cycle aging. Total costs are the sum of energy and degradation costs.

The comparison between the years 2013 and 2017 shows that total costs are slightly lower in 2017 which is simply a consequence of the lower electricity price in this year. As energy costs only contribute approx. 20–40 % to total costs, the electricity price difference of 17.7 %, however, only manifests itself in relatively small savings of about 3.7–7.7 %. As in all scenarios the same amount of energy is charged, energy costs are the same for all scenarios within the same year; slight deviations are a result of rounding inaccuracies.

The major cost factor is found to be battery degradation with degradation costs accounting for approx. 57–82 % of total costs. The major contribution to degradation stems from calendar aging as indicated by the idling costs which amount to 73–86 % of degradation costs. This is due to the fact that the average vehicle only consumes about 3 kWh of energy per day so that not much cycling is performed. Most of the depreciation is therefore not related to the creation of value. Since the uncoordinated charging strategy recharges the battery to the target SOC as soon as a grid connection is established, the battery is generally stored at a high SOC which according to the discussion in Section 5.3 is unbeneficial with regard to calendar degradation. Idling costs and therefore degradation costs are higher at higher target SOC values. Reducing the target SOC from 0.8 to 0.5 reduces degradation costs by approx. 42 % which in absolute numbers is of the same order as the total amount paid for electricity during the investigated period of time.

Doubling battery capacity almost doubles degradation cost. A factor of 2 stems from the doubling of calendar aging costs since twice the battery value depreciates during the same lifetime. Cycle aging costs decrease slightly because cycling the same amount of energy corresponds to a smaller SOC swing for a bigger battery which results in less cycle aging. For the typical PEV owner with a moderate daily mileage, battery degradation, and more specifically calendar degradation, is therefore the driving cost factor. From the considerations presented above, it becomes clear that a charging strategy which consistently maintains a high SOC has a particularly undesirable effect on PEV operating costs. High costs also result from an overdimensioned battery capacity which buys more flexibility for the PEV owner at a relatively high price. Notable savings can therefore be expected from employing a charging strategy which optimizes the charging schedule based on expected energy needs and from reducing battery capacity to the required minimum.

5.5.3.2 Smart Charging

The results of the smart charging strategy are shown in Table 5.7 with the numbers in parentheses indicating the relative difference to the uncoordinated charging scenario with a target SOC of 0.8. As these numbers show, total costs can be reduced by 40–46 % in both years when employing the smart charging strategy. This is a notable difference which is mainly a result of savings in battery degradation costs which are reduced by approx. 55 %. This is primarily due to the fact that the smart charging strategy generally maintains a low SOC level until shortly before a trip starts. Energy cost savings up to 10 % are also observed since the scheduler is able to make use of

Table 5.7: Costs of the smart charging strategy for one week. The numbers in parentheses indicate the change compared to the uncoordinated charging scenario with a minimum SOC of 0.2 and a target SOC of 0.8.

Year	2013		2017	
Batt. cap. [kWh]	10	20	10	20
Energy costs [$]	3.7	3.8	3.1	3.2
	(−10 %)	(−5 %)	(−9 %)	(−3 %)
Degrad. costs [$]	4.0	6.7	4.0	6.7
	(−54 %)	(−56 %)	(−54 %)	(−56 %)
Idling costs [$]	2.8	5.4	2.8	5.6
	(−57 %)	(−57 %)	(−57 %)	(−57 %)
Total costs [$]	7.7	10.5	7.1	9.9
	(−40 %)	(−45 %)	(−41 %)	(−46 %)

lower electricity prices. These are, however, moderate in comparison with degradation cost savings.

Figure 5.12 shows the share of agents according to their relative savings for the year 2017 and a battery capacity of 10 kWh. It can be seen that most agents save between 35–55 % as compared to the uncoordinated charging strategy. Lower savings are only observed in a very small number of cases with large mileages where energy is the primary cost driver. Since these vehicles have shorter parking times, the scheduler only has limited flexibility for charging so that the fluctuation of prices cannot be exploited. Higher savings may occur at very low mileages which only apply to a negligibly small number of agents.

5.5.3.3 Energy Arbitrage

The costs resulting from the arbitrage strategy are presented in Table 5.8. Compared to the uncoordinated charging strategy, total costs are reduced by 46–55 %. In the high-price year 2013 this yields another 6–10 percentage points in savings compared to the smart charging strategy while in the low price year only a maximum reduction

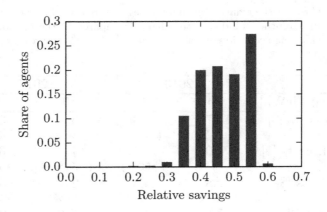

Figure 5.12: Share of agents as a function of their relative savings with smart
charging in the year 2017 with a battery capacity of 10 kWh.

by 1 percentage point is achieved. The additional revenue in the year
2017 is therefore negligible and is unlikely to cover transaction costs.
In the high-price year 2013, an additional benefit can be attained
through energy arbitrage, however, compared to the gains achieved
through smart charging only, this effect is very moderate.

Table 5.8: Costs of the arbitrage strategy for one week. The numbers in parentheses indicate the change compared to the uncoordinated charging
scenario with a minimum SOC of 0.2 and a target SOC of 0.8.

Year	2013		2017	
Batt. cap. [kWh]	10	20	10	20
Energy costs [$]	4.9	6.6	3.3	3.7
Degrad. costs [$]	4.7	8.1	4.1	7
Idling costs [$]	2.9	5.7	2.8	5.5
Revenue [$]	2.7	6.1	0.4	1
Total costs [$]	6.9	8.6	7.0	9.7
	(−46 %)	(−55 %)	(−42 %)	(−47 %)

5.5.3.4 Comparison of Plug-In Electric Vehicle and Internal Combustion Engine Vehicle Costs

In order to put the computed figures into relation with the currently predominant propulsion technology, a simple cost comparison with ICEV costs is performed. The comparison is limited to the two most distinctive features of ICEVs and PEVs which are their fuel economy due to different energy carriers and the energy storage technology. With regard to these two factors PEVs and ICEVs have opposite economic advantages. PEVs are superior to ICEVs in terms of fuel economy as the following numbers illustrate: While an electric BMW i3 has a rated power consumption of 12.9 kWh per 100 km, a gasoline powered BMW 116i is rated at 5.4 l per 100 km. This is an energy equivalent of approx. 51 kWh per 100 km given a gasoline energy content of 9.44 kWh per liter. Assuming electricity is generated in a modern gas fired power plant with an efficiency of 60 %, the ICEV consumes almost 2.4 times as much primary energy as the PEV. With regard to the energy storage technology, the advantage is on the side of the ICEV since it does not require an expensive battery. Given the capacity of 20 kWh and a price of $ 200 per kWh, this makes a difference of $ 4 000.

As by the time of writing CityMoS traffic does not contain any ICEV models, the simulated driving distances are simply multiplied by the specific fuel consumption listed in Section 5.5.1. Using this number, the average operating cost of an ICEV is $ 17.5 per week. This is slightly less than in the case of a PEV using the uncoordinated charging strategy with an SOC target of 0.8, a 20 kWh battery and 2017 data where weekly costs sum up to $ 18.5. If the smart charging strategy is employed, weekly costs of the PEV drop to $ 9.9 so that in this case the PEV is considerably more competitive with regard to the considered cost aspects.

Due to the trade-off between fuel economy and battery costs, the PEV's economic competitiveness increases with the daily driving distance. This is illustrated in Figure 5.13a which compares the weekly

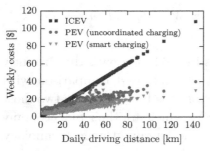 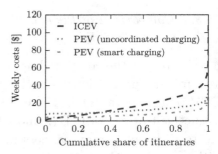

(a) Costs as a function of driving dis-   (b) Costs by cumulative share of
    tance.   itineraries.

Figure 5.13: Comparison of ICEV and PEV operating costs for different
        itineraries with uncoordinated charging and smart charging for a
        battery capacity of 10 kWh.

costs of the uncoordinated charging strategy, the smart charging strategy and the ICEV for all agents as a function of driving distance.
The data represents the scenarios with a 10 kWh battery and a target SOC of 0.8. Both uncoordinated and smart charging begin at a
higher cost level than the ICEV since calendar degradation is present
in any case. The ICEV costs are strictly linear over the distance by
definition while the PEV costs lie within a band with a similar slope
(0.2 for uncoordinated charging and 0.24 for smart charging). Outliers in Figure 5.13a are in most cases a result of single trips with
long distances where the scheduler needs to charge to a high SOC
which accelerates degradation. As shown in Figure 5.13b, the smart
charging strategy achieves a break even after 10 % of all itineraries
while the uncoordinated charging strategy hits the break even point
at 32 % of all agents.

Considering that under these conditions cost parity is achieved for
68–90 % of all agents, this results in a large percentage of users who
would economically benefit from a PEV, given the very small battery
capacity of 10 kWh. The share, however, decreases quickly as the battery capacity increases since degradation costs almost scale linearly
with battery capacity. As shown in Figure 5.14 for the uncoordi-

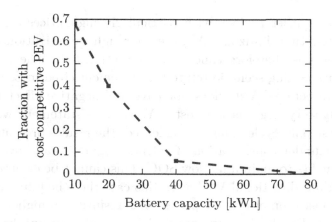

Figure 5.14: Share of agents with cost-competitiveness of PEVs with ICEVs
for different battery capacities in the uncoordinated charging
scenario.

nated charging strategy, in the case of a battery capacity of 20 kWh,
only 40 % of the population economically benefit from a PEV. This
decreases further to 5 % for 40 kWh and to 0 % for 80 kWh.

While this is only a crude comparison of which the pitfalls are
discussed in Section 5.5.4, it still demonstrates that a smart charging
strategy and the influence of calendar degradation can play a vital
role with regard to the economic competitivity of PEVs.

5.5.4 Conclusion

In summary, the presented results show that a smart charging strategy can lead to considerable cost savings for PEV owners. As opposed
to the common conception of the importance of charging at low electricity prices, the investigations show that the greatest savings can
be achieved by optimizing the charging with regard to battery degradation. In this context, the often neglected contribution of calendar
degradation plays the most important role. Relatively small benefits
are observed for additionally conducting energy arbitrage by means
of bi-directional V2G. This may, however, also be a consequence of

the lack of outlier periods with particularly high prices during the investigated time horizon. A more thorough investigation of arbitrage profits is therefore conducted in Section 5.6. The results of the smart charging scenario refute the statement cited in Section 2.6 that uni-directional V2G does not have any negative effect with regard to battery degradation costs. While this statement would be true in case only cycle aging is considered, the results presented here show that under consideration of calendar aging it can have considerable benefits to maintain a low SOC. This cannot be ensured when providing uni-directional V2G as in this case the scheduling decision would be based on electricity prices, thus resulting in higher average SOC levels. This leads to the conclusion that also for uni-directional V2G battery degradation has to be considered in order to trade off additional degradation costs with V2G revenues.

The mobility patterns used for this study are simulated using CityMoS Traffic with 20 000 agents. This number is well below the value of the private car population of approx. 540 000 in Singapore [205]. The reason for this is that at the time of conducting the experiments the capability of the simulation was limited to this value. This is not an issue in terms of a representative sample of the population, nevertheless, this limitation may lead to an underestimation of energy consumption. This is because it reduces traffic density which in turn reduces the number of acceleration and deceleration processes thus lowering energy consumption. Another factor which may lead to a further underestimation of energy consumption is that only a small number of traffic lights are implemented. Greater numbers of traffic lights, however, would result in more interruptions of traffic flows, thus increasing energy consumption. These aspects may be reflected in the rather low average energy consumption of 12.8 kWh per 100 km which corresponds to the rated energy consumption of a BMW i3 with 12.9 kWh per 100 km. The U.S. *Environmental Protection Agency* (EPA) rates the same vehicle with 15.3 kWh for city traffic which can be assumed to be a more realistic number [206].

When applying the simulation results to planning scenarios for Singapore, this discrepancy of approx. 20 % would have to be taken into consideration. Nevertheless, even if actual energy consumption was increased by 50 %, 95 % of all agents would still remain below a daily energy consumption of slightly more than 10 kWh.

It is further noted that the daily mileage of 23.1 km which is computed using CityMoS Traffic is not well in accordance with other data provided by the *Land Transport Authority* (LTA) of Singapore [205]. According to this data, the annual mileage of private cars in 2014 was 17 500 km corresponding to approx. 48 km per day. As this data stems from mileage surveys based on mandatory periodic vehicle inspections, it also contains trips outside of Singapore. This may be a reason for the discrepancy, however, it may have to be taken into consideration that the simulation could be underestimating total mileage. In this case, the results would have to be scaled accordingly. As this work does not primarily aim to establish planning scenarios for Singapore, these issues are of relatively low importance in this context. For providing policy or infrastructure planning advice, these aspects, however, would require further investigation.

With regard to the smart charging and arbitrage strategy, it can be objected that the lookahead of 7 periods does not achieve maximum savings. As shown in Section 5.4, a lookahead beyond 7 periods, however, increases profits by less than 2 % in an intraday arbitrage scenario which can be neglected. Furthermore, large lookaheads come with greater forecasting inaccuracies which would therefore not necessarily result in a better outcome.

This study also does not exploit any potential for interday arbitrage which may exist as a consequence of different load patterns between different days, especially between working days and weekends. As argued in [116], however, intraday arbitrage is responsible for most value creation so that no major gains from interday arbitrage should be expected.

The comparison of ICEVs and PEVs does not aim to provide a thorough analysis on the competitiveness of ICEVs and PEVs but instead isolates the effect of the most important cost drivers. A more complete consideration would involve a comprehensive analysis on a TCO basis. This would include the value depreciation of other components, insurance and taxes, costs for maintenance and repair as well as expenses for charging installations. The computed numbers should therefore not be generalized without taking these additional cost factors into account. An important conclusion from this investigation which common TCO analyses tend to ignore, however, is that not only external cost parameters such as battery prices but also the employed charging strategy can have a considerable impact on the competitiveness of PEVs and ICEVs. In any case, due to the large number of parameters, the cost considerations presented here as well as those in other studies can only be considered as snapshots covering a small number of scenarios. As it is shown in this study, cost factors can be considerably tweaked by adapting battery or charging strategy parameters so that these cost considerations need to be tailored to a particular use case scenario. The presented framework can support these investigations by providing the models required to assess battery related depreciation in accordance with the respective operating conditions.

5.6 Economic Viability of Battery Energy Storage Systems in Different Markets

In this section, the economic viability of BESSs for energy arbitrage in different real-time markets is investigated. The objective of this study is to provide an insight into conditions under which energy arbitrage using small-scale BESSs can be profitable. The investigation further illustrates the role of battery degradation consideration and electricity price forecasts, the relevance of the temporal and spatial resolution of price data as well as the effect of cost savings due to decreasing battery prices.

5.6.1 Markets

The markets under consideration are the Ontario market operated by the *Independent Electricity System Operator* (IESO) [207], the New York market operated by the *New York Independent System Operator* (NYISO) [208], the NEMS [209] (Singapore) and the PJM [210] (eastern area of the U.S.). The relevant characteristics of the investigated markets are very similar and in accordance with the introduction into electricity markets presented in Section 2.3. The submarkets under consideration are the real-time generation markets.

The market operators provide data on NMPs and ZMPs which are computed for distinct nodes or zones based on current and projected capacity utilization. All market operators provide spatially aggregated prices consisting of the weighted average LMPs. Spatially and temporally averaged prices are typically used as wholesale prices paid by contestable consumers in the entire market area. Time resolutions of these prices are 60 min (IESO, NYISO, PJM) and 30 min (NEMS).

5.6.2 Parameters

The results presented in Section 5.6.3.1 build on the aggregation level of wholesale prices while in Sections 5.6.3.2 and 5.6.3.3 zonal and nodal prices in the NYISO area with different time resolutions are also considered. The used data includes zonal prices representing subregions of the market as well as nodal prices for 508 generators. The time period under consideration covers the years 2008–2016.

For scheduling, the lookahead is set to 7 periods in accordance with the results in Section 5.4. As in this investigation perfect forecasts based on cleared prices are assumed, the computed profits define an upper boundary. It can, however, be assumed that even with imperfect price forecasts approx. 85 % of the revenues calculated under these simplified conditions could be captured [116]. This argument is also supported in [211] based on work presented in [212–215].

For the battery price, scenarios of $0 per kWh, $150 per kWh and $300 per kWh are investigated. The zero battery cost scenario serves as a baseline illustrating the implications of not explicitly considering degradation costs for scheduling. According to [57], $150 per kWh is considered a break even cost at which PEVs are expected to become a competitive mass market product. The same source considers the price of $300 per kWh as a realistic number which can currently be achieved for PEV and home batteries. The price range therefore considers an optimistic and a realistic scenario. The SOC discretization required by the DP solver is set to 0.05 which is a sufficiently good compromise between accuracy and performance as shown in Section 5.4.

5.6.3 Results

This section presents the results for BESS profitability in various scenarios considering wholesale prices, nodal and zonal prices and different EOL criteria.

5.6.3.1 Wholesale Market Comparison

To demonstrate the importance of appropriately accounting for battery degradation, the profits in the various markets are first computed without the consideration of calendar aging. The corresponding results are shown in Figure 5.15.

As the scheduling algorithm does not account for degradation costs, in this case the scheduler acts as a bang-bang controller which either keeps the battery idle or charges/discharges at maximum power. This results in heavy cycling so that the battery is subject to fast capacity fade. The implications can be quantified by computing the real degradation related to these schedules using the full degradation model. This leads to the conclusion that under the obtained cycling regime, the average normalized capacity in the various markets would decrease to 0.69–0.84 in one year. Taking the EOL criterion

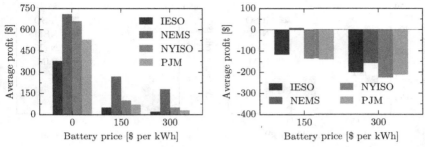

(a) Average annual profits without considering calendar aging.

(b) Real average annual profits including the consideration of calendar and cycle aging based on the schedules computed for Figure 5.15a.

Figure 5.15: Average annual profits with a lookahead of 7.

$\Gamma_{EOL} = 0.8$ for PEVs, this could result in up to 1.5 battery replacements per year which is in line with the findings in [1] and [2].

The central and rightmost sets of bars show the scenarios where the battery degradation model considering cycle aging is included in the optimization. For a battery price of $ 150 per kWh, average annual profits range from $ 50 (IESO) to $ 270 (NEMS). While these numbers are considerably lower than in the case without consideration of battery costs, they indicate profits with cycle degradation costs factored in. Normalized capacities in these cases only decrease to a minimum of 0.95 over the course of a year. The degradation model thus ensures that energy is only cycled under conditions where revenues exceed cycle degradation costs. With battery prices of $ 300 per kWh, profits decrease to $ 20 (IESO) to $ 180 (NEMS). Capacity losses further decrease slightly compared to the scenario with a battery price of $ 150 per kWh to a worst case of 0.96, indicating that in this case even smaller amounts of energy are cycled.

The implications of neglecting calendar aging become obvious when calculating the real profits related to the schedules computed for Figure 5.15a. This is done by applying the full degradation model including calendar aging on the charge/discharge cycles behind Fig-

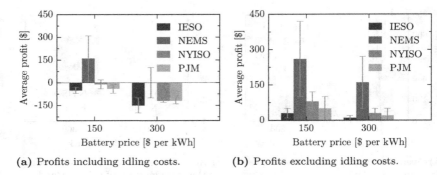

(a) Profits including idling costs.   **(b)** Profits excluding idling costs.

Figure 5.16: Average annual profits computed considering calendar and cycle
aging with a lookahead of 7.

ure 5.15a. The result of this calculation is shown in Figure 5.15b. As
the way of calculating degradation is irrelevant if battery prices are
0, the leftmost set of bars in Figure 5.15a would remain unaffected
which is why they are not shown. For battery prices of $ 150 per kWh
and $ 300 per kWh, however, it is observed that the outcome ranges
from a profit of $ 10 up to a loss of $ 225 so that virtually all markets
are loss-making.

The situation looks different in case calendar aging is explicitly
considered when computing optimal schedules as shown in Figure 5.16.
In Figure 5.16a, it can be seen that under these conditions arbitrage can be profitable at NEMS, given battery prices are below
$ 300 per kWh. At a battery price of $ 150 per kWh, the annual profit
in this market amounts to $ 160. All other markets still yield losses
which are, however, notably reduced compared to the results shown
in Figure 5.15a. In these markets, investing in a BESS for the sole
purpose of energy arbitrage is therefore not profitable under the investigated conditions.

If a BESS has been purchased for a reason other than energy arbitrage, e.g., as a PEV or as a residential BESS to buffer PV generation,
this calculation looks different. In this case, the costs resulting from
calendar degradation while idling the battery can be considered as

sunk costs so that they do not have to be subtracted from arbitrage profits. Figure 5.16b shows the profits excluding these idling costs. In this case, it can be seen that profits can be made in all four markets. Profits are highest at the NEMS with $260 for the lower battery price and $160 in the case of the more expensive battery. The second most profitable market is the NYISO with average annual profits of up to $80 for the low cost battery. While the amount computed for the NEMS may be sufficiently high as a basis for a business case, profits in any of the other markets are assumed to be too low to justify transaction costs.

5.6.3.2 Spatial Resolution

Profitability of BESSs may be increased by selecting appropriate locations to benefit from NMPs or ZMPs. The effect of identifying the most profitable locations for BESS grid integration is investigated by computing profits in the different load zones of the NYISO area for the years 2008–2016 with a lookahead of 7 using the full degradation model. The results for a battery price of $150 per kWh listed in Table 5.9 show that the zonal profits averaged over these years range from $0 (*Genesee*) to $137 (*Long Island*). This is well above the case with a uniform market price. In all zones, the spread between minimum and maximum profits over the different years is fairly large with a maximum annual loss of $49 (*Genesee*) and a maximum annual profit of $306 (*Long Island*). While there is no average loss incurring in any of the zones, *Long Island* is the only one which is profitable throughout the entire time horizon. In this case, the profit corresponds to an average annual ROI of 4.6% which may be sufficiently high to justify the purchase even for the purpose of arbitrage only. In all other zones, however, profits are most likely too low to be financially attractive.

This situation is different when energy arbitrage is not the primary reason for the investment so that idling costs can be excluded. In this case, average annual profits in a range between $98 and $269

Table 5.9: Annual profits in the different NYISO zones for the years 2008–2016 for a battery price of $ 150 per kWh.

| | Annual profits [$] | | | | | |
| | Including idling costs | | | Excluding idling costs | | |
Zone	Avg.	Min.	Max.	Avg.	Min.	Max.
Capital	18	-40	131	123	45	293
Central	5	-40	133	104	43	291
Dunwoodie	46	-24	131	151	68	278
Genesee	0	-49	129	98	38	284
Hudson Valley	33	-28	110	136	63	253
Long Island	137	48	306	269	169	463
Mohawk Valley	7	-43	133	107	46	297
Millwood	43	-24	125	148	68	271
New York City	52	-14	133	163	80	293
North	21	-43	169	124	49	337
West	45	-37	128	148	45	260

are made which is likely to yield sufficiently high returns even after deducting possible transaction costs.

At the higher battery price of $ 300 per kWh as shown in Table 5.10, losses incur in all zones if idling costs are considered. In the case where idling costs are excluded, moderate average profits in a range between $ 47 and $ 145 per year are possible. This indicates that at current battery prices, most zones would not be sufficiently attractive for energy arbitrage.

While profits vary across the different years due to varying price levels, the general rank of individual zones in terms of their profitability changes little over time. This is illustrated in Figure 5.17 which shows that the profits in the individual zones largely follow the general trend. This indicates that the most profitable locations could be selected based on historical data in order to maximize profitability.

Table 5.10: Annual profits in the different NYISO zones for the years 2008–2016 for a battery price of $ 300 per kWh.

| | Annual profits [$] | | | | | |
| | Including idling costs | | | Excluding idling costs | | |
Zone	Avg.	Min.	Max.	Avg.	Min.	Max.
Capital	-106	-132	-74	52	16	118
Central	-113	-138	-76	46	16	127
Dunwoodie	-81	-123	-38	76	28	134
Genesee	-117	-152	-82	43	10	125
Hudson Valley	-90	-127	-52	66	22	118
Long Island	-55	-101	37	145	64	251
Mohawk Valley	-113	-145	-77	47	13	126
Millwood	-89	-124	-49	87	28	247
New York City	-81	-125	-49	78	27	125
North	-114	-145	-63	59	15	151
West	-92	-130	-45	72	17	136

The influence of the location becomes even more eminent when considering prices at a nodal instead of a zonal level. For this pur-

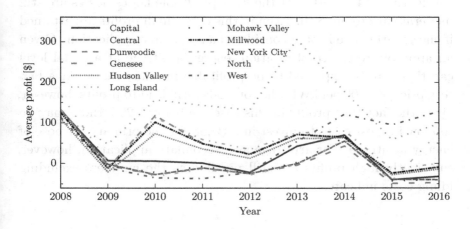

Figure 5.17: Average annual profits over time in the different NYISO zones for a battery price of $ 150 per kWh.

Table 5.11: Three most and least profitable NYISO nodes in 2015 for battery
prices of $ 150 and $ 300 per kWh.

	Annual profits [$]			
	$ 150 per kWh		$ 300 per kWh	
	Incl.	Excl.	Incl.	Excl.
	idling	idling	idling	idling
Node	costs	costs	costs	costs
HUNTLEY____68	480	618	186	395
HUNTLEY____67	480	618	186	395
BARRETT____1	387	539	101	310
GINNA_____	-54	32	-141	8
O.H._GEN_BRUCE	-56	29	-143	7
KINTIGH_____	-64	23	-150	5

pose, the profits attainable in the 508 generator nodes of the NYISO
market in 2015 are computed. The profits for the three most profitable and the three least profitable nodes are listed in Table 5.11.
For a battery price of $ 150 per kWh, the two top nodes yield a profit
of $ 480 when considering idling costs and a profit of $ 618 without
considering idling costs. At the least profitable nodes, losses are still
moderate at around $ 50. Even in this case, small profits are attained
if energy arbitrage is only considered a secondary application. Given
an appropriate choice of location, energy arbitrage at a nodal level
can therefore be expected to be profitable in many cases. At a battery price of $ 300 per kWh, the top nodes still lead to profits, however,
given the fact that profits in this case are below $ 200, these may be
too low in most cases to invest in a BESS for the primary purpose of
energy arbitrage. For arbitrage as a secondary application, however,
sufficiently high profits are still possible as the numbers excluding
idling costs indicate.

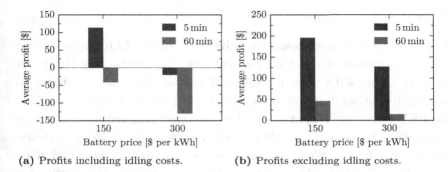

(a) Profits including idling costs. (b) Profits excluding idling costs.

Figure 5.18: Arbitrage profits for time resolutions of 5 min and 60 min with a lookahead of 7 for the year 2015.

5.6.3.3 Temporal Resolution

Profitability can also be increased by leveraging on a higher temporal resolution which provides greater arbitrage potential. This is investigated by computing profits using the zonal average prices for the year 2015 with a 5 min resolution and by comparing the results to the case of the 60 min resolution. As shown in Figure 5.18a for a battery price of \$ 150 per kWh, this yields a loss of approx. \$ 40 at a time resolution of 60 min. At the higher time resolution of 5 min, however, a profit of approx. \$ 130 can be achieved. At a battery price of \$ 300 per kWh, losses are made in both cases but these are increased by a factor of 6.5 for the 60 min resolution. (The results for the 60 min resolution in Figure 5.18a differ slightly from those in Figure 5.16 because the 60 min prices used here are simple temporal averages of the 5 min prices whereas the results in Figure 5.18a are based on weighted averages). When excluding idling costs as shown in Figure 5.18b, profits are made in all cases with an increase by factors of approx. 4–9 for a time resolution of 5 min. The results show that given current battery prices and lifetimes, the time resolution of price data can play a major role regarding profitable operation.

5.6.3.4 EOL Criterion

In Section 5.4, it was shown that the value of the EOL capacity can have a significant impact on profitability. For stationary BESSs, a value of $\Gamma_{EOL} = 0.8$ is very conservative since stationary systems do not exhibit strict requirements with regard to energy density. Also for PEV batteries, recent research states that this criterion is overly pessimistic by concluding that given a sufficiently developed charging infrastructure, 85 % of travel needs can still be satisfied with $\Gamma_{EOL} = 0.57$ [197]. This section therefore presents the profits for the same setting as in Figure 5.16 for $\Gamma_{EOL} = 0.6$. The results are shown in Figure 5.19 with Figure 5.19a depicting the profits including idling costs and Figure 5.19b showing the profits excluding idling costs. As opposed to the case with $\Gamma_{EOL} = 0.8$, it can be seen in Figure 5.19a that profits are made in all markets at a battery price of \$ 150 per kWh. For a battery price of \$ 300 per kWh, three markets are profitable while no profits were made for $\Gamma_{EOL} = 0.8$. If idling costs are excluded as shown in Figure 5.19b, all markets can be considered sufficiently profitable for a valid business case except for the IESO at high battery prices. In accordance with the discussion of the influence of the EOL criterion on battery degradation costs in Section 5.3, profits more than double when doubling $\Delta\Gamma_{EOL}$. These results show that particularly for stationary applications where lower EOL capacities can be accepted, and in particular for second life batteries, energy arbitrage can be sufficiently profitable.

5.6.4 Conclusion

The presented results show the relevance of appropriate consideration of battery degradation and underline that the failure to account for this aspect is almost certain to result in financial losses. The various investigations also show that under the given parameters, BESSs are at the edge of profitable operation for energy arbitrage. Based on the obtained results, it can be concluded that under unbeneficial framework conditions, i.e., battery prices of around \$ 300 per kWh, strict

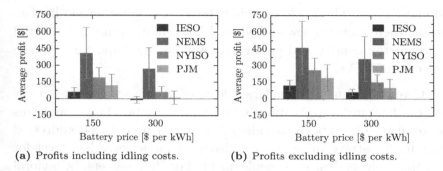

(a) Profits including idling costs.   **(b)** Profits excluding idling costs.

Figure 5.19: Average annual profits for $\Gamma_{EOL} = 0.6$.

EOL criteria and access to uniform market prices only, BESSs for the primary purpose of energy arbitrage cannot be expected to be profitable.

Under favorable conditions, however, profitable operation is possible. This is particularly true for applications where the BESS serves a different primary purpose so that the expenses for purchasing the battery can be considered as sunk costs. In this case, financial losses cannot occur since the BESS controller keeps the battery idle if cycling energy would be loss-making. This, however, requires a degradation-aware scheduling technique capable of accounting for this trade-off. Under these conditions, arbitrage profits need to be sufficiently high to cover transaction costs. At uniform market prices, this can only be expected to hold for the NEMS if the EOL is assumed to be reached at a remaining capacity of 0.8. Relaxing this condition to 0.6, however, can result in profitable operation in virtually all markets, even at battery prices of $ 300 per kWh. With access to zonal prices, all NYISO zones could also be considered profitable under these conditions if battery prices are around $ 150 per kWh.

To justify the investment into a BESS for the sole purpose of energy arbitrage, framework conditions have to be particularly beneficial. With access to nodal electricity prices at high temporal resolutions and low battery prices, sufficiently high profits could be attained in many cases. This could especially be a use case for second life batteries where costs are low and where EOL conditions are not as strict since energy density and power capacity are less important than in the case of PEVs.

Given the generally moderate level of profits at current battery prices, those niche applications with particularly beneficial conditions according to the above considerations would need to be addressed first until battery prices have fallen to a sufficiently low level for location-independent mass deployment. For PEV owners, behavioral aspects including range anxiety may pose a barrier for participating in power grid services which may not be countered by sufficiently high profits. Therefore, stationary applications with residential batteries combined with PV modules or second life batteries are considered more likely applications. Any of these implementations, however, require appropriate schemes for battery owners to participate in real-time and day-ahead markets and to get access to nodal or zonal prices. For commonly discussed approaches such as the involvement of a VPP operator, sufficiently high margins to cover the additional costs for this entity need to be considered in the cost assessment. It also needs to be acknowledged that the battery prices as used in this work represent manufacturing costs so that end-consumer prices are typically higher.

5.7 Conclusion

This chapter presents several applications of the scheduling framework which was introduced in Chapters 3 and 4. One focus is set on the investigation of the relevance of appropriately considering battery degradation for optimal scheduling. It is shown that degradation costs considerably vary across two orders of magnitude depending on operating conditions. This leads to the conclusion that the assumption of an average constant cost taken in a large body of research results in far sub-optimal profits or financial losses. Also for approximate estimations of the profitability of BESSs in different energy markets, the commonly used assumption of average degradation costs cannot be considered appropriate since setting an average cost value is subject to high uncertainty and would require good knowledge on the expected cycling conditions. The results further show that in a general case neither the processes related to cycle aging nor those resulting from calendar aging can be neglected. In contrast, the contribution of either degradation process is again highly dependent on the operating conditions including internal factors such as the cycling regime and external aspects such as ambient temperature. Especially in cases where the battery is not subject to heavy cycling and under circumstances where the controller would otherwise maintain high SOC levels, the consideration of calendar aging is inevitable for correct cost estimates. In summary, these observations lead to the conclusion that a framework as presented in this work can significantly improve the scheduling of BESSs and support the preparation of substantiated profitability analyses.

For PEVs, the comparison of an uncoordinated charging strategy, which does neither account for electricity price fluctuations nor for battery degradation, with smart charging and energy arbitrage clearly shows the financial disadvantages of uncoordinated charging. In contrast to the common conception, however, electricity cost savings resulting from price-responsive charging are not the biggest contributor to overall savings. Instead, the greatest contribution to savings results from the reduction of calendar degradation costs by maintaining low SOC levels.

In this context, it is also shown that the tendency of oversizing batteries comes at a relatively high cost as the unused capacity is naturally subject to calendar degradation. The conception that larger battery capacities could even result in lower total costs due to smaller SOC swings for given mobility patterns can therefore not be confirmed. Instead, a closest possible fit of battery capacity to the actual requirements is considered most economical. This is also illustrated by the observation that at a battery capacity of 10 kWh, approx. 68 % of Singapore drivers could benefit from a PEV as compared to an ICEV, even with an uncoordinated charging strategy. At the same time, this percentage decreases to a mere 5 % for 40 kWh of capacity. The investigation further shows that the additional gain from energy arbitrage as compared to uni-directional smart charging under the considered conditions is moderate, especially in years with low electricity prices.

The market comparison for stationary BESSs arrives at a similar result with regard to arbitrage profitability for current battery price levels if the remuneration is based on uniform market prices. In this case, the investment in a BESS for the sole purpose of energy arbitrage yields a negative ROI in most cases. Moderate profits can be achieved if energy arbitrage is only considered a secondary application of the BESS so that depreciation costs which do not directly result from arbitrage-related energy throughput can be considered as sunk costs. Higher profits can be achieved if remuneration is based on nodal or zonal prices or prices with higher temporal resolution. It is therefore concluded that beneficial conditions such as moderate battery prices, an appropriate choice of location and a scheduling mechanism designed to maximize profits can indeed lead to a profitable operation with sufficiently high ROI. As conclusions on profitable operation depend on a significant number of parameters, a framework like the one developed in this work should be deployed for conducting substantiated profitability analyses for the particular use case under consideration prior to making any investment decision.

6 Conclusion

6.1 Summary

This section summarizes this work with an overview of the developed methodology in Section 6.1.1 and of the obtained results in Section 6.1.2.

6.1.1 Methodology

This work presents a framework for optimal charge scheduling of grid integrated ESSs. As described in detail in Chapter 3, the framework is designed in a modular way to allow easy modification, extension and addition of individual models in order to account for the large variety of possible scenarios. This permits to tailor the simulation environment to the actual problem to be addressed in order to identify optimal system design parameters and to compare different scenarios with each other. While the focus of the presented work is on BESSs, the modular approach allows extending the framework's capability to virtually any kind of ESS. Furthermore, extended market models, price forecasting mechanisms or additional load models can be integrated. By strictly separating source code, data and configuration, the framework allows the systematic and automatic investigation of different scenarios with a variety of parameter and model combinations.

The current implementation particularly aims to assess the profitability of energy arbitrage using BESSs in electricity markets. For this purpose, the framework comprises a scheduling approach which allows optimizing the charging and dispatching of energy by considering real-time electricity prices, price forecasts, availability constraints and battery degradation. The methodology represents an MPC approach implemented as an indirect control scheme where the BESS responds to price signals in order to maximize its utility. These price signals may be generated by the PSO or a VPP operator as part of a DSM scheme aiming to optimize the load flow in the power system. While the price signals are provided by an external entity thus considering the BESS as a price taker, entire control remains on the side of the BESS as a consequence of the indirect control scheme. This is assumed to be particularly relevant with regard to PEVs and distributed BESSs. In these cases, the battery generally serves another primary purpose such as providing a sufficient amount of energy for driving or for buffering PV generation while serving residential loads. Computing optimal schedules requires detailed information on mobility patterns or residential load patterns. These are best known on the side of the individual system and system owners might be hesitant to share this information for privacy reasons. Furthermore, optimal battery operation requires extensive knowledge on the BESS itself, especially on the battery's degradation behavior, which can be most easily achieved by a controller which is part of the system.

In order to appropriately account for battery constraints and battery degradation, the framework comes with a battery model and a battery degradation monetization model as described in Chapter 4. The first one establishes a relation between charge and energy throughput and ensures that the battery is charged and discharged in an appropriate way. This includes the consideration of CC and CV charging modes which are common for Li-ion batteries. Considering battery degradation as part of the scheduling mechanism is identified to be the most crucial factor to ensure profitable operation of a BESS. The framework therefore implements a module for quantifying and monetizing battery degradation in order to optimally control battery wear. The methodology for monetizing battery degradation is formulated in a generic way, thus serving as a blueprint for the integration of customized degradation models matching the properties of the cell of interest. The deployed model is capable of considering both calendar and cycle degradation for capacity and resistance related aging and allows tracking the cell's SOH with regard to these quantities over time. The marginal degradation costs computed by the degradation model are used as part of the scheduler's utility function, thus allowing to trade off possible revenues with battery degradation costs. This ensures that the BESS is operated in a cost-optimal way.

The developed methodology may be used in conjunction with the simulation frameworks CityMoS Traffic and CityMoS Power in order to extend the range of possible applications with regard to large-scale PEV charging and its impact on a city's power system.

6.1.2 Results

The framework was deployed to provide an insight into the relevance of battery degradation for optimal scheduling, to assess the financial benefits of degradation-aware smart charging of PEVs in a Singapore context and to investigate the profitability of energy arbitrage using BESSs in different markets. These applications are presented in Chapter 5.

The investigation with regard to battery degradation shows that the appropriate consideration of battery wear has a crucial impact on optimal system operation. The results indicate that depending on operating conditions, degradation costs per unit of cycled energy can easily spread across two orders of magnitude. On the one hand, these numbers show that the assumption of constant degradation costs, which underlies a great number of profitability assessments, is very problematic since this would require a precise understanding of what the average cycling regime would be. On the other hand, the large spread of costs also reveals considerable optimization potential since the cycling conditions are within the controller's scope of influence. Controlling the charging and dispatching with regard to this cost aspect can therefore make the difference between loss-making and profitable operation.

The analysis also shows that neither the calendar nor the cycle contribution to degradation can generally be neglected. In the investigated cases, the unit cost corresponding to a full charge or discharge was computed to be $0.21 per kWh with considering cycle aging only as compared to $0.35 per kWh with considering both calendar and cycle aging. This difference can increase further if the battery is not subject to heavy cycling and if there is a tendency towards high SOC levels. Both of these aspects often apply to PEVs and to a potentially somewhat lesser extent to residential BESSs. In these cases, calendar aging can also be the dominant cost driver. Neglecting the calendar contribution to degradation as done often can therefore lead to a considerable distortion of results.

Using the developed framework, different charging strategies for PEVs were compared in a Singapore context. This investigation was performed in conjunction with the traffic simulation CityMoS Traffic which served for assessing the charging requirements of Singapore drivers. The results show that smart charging strategies can lead to notable savings for the PEV owner as compared to uncoordinated charging. In the investigated cases, savings of up to 46 % with the smart charging strategy compared to uncoordinated charging were observed. Against the more common conception, these savings are, however, not found to be primarily a consequence of electricity cost savings. Instead, more than 90 % of the savings can be attributed to reduced battery degradation, specifically as a consequence of reducing calendar aging. This is mainly due to the fact that the smart charging strategy only charges the battery up to a level required to satisfy the driving needs for an upcoming trip. Implementing this strategy in reality, however, presumes knowledge on a driver's mobility pattern and requires the driver to accept a battery which is not fully charged. A lighter version of this strategy would be to delay charging until shortly before a trip starts so that the battery is stored for a shorter time at a high SOC level.

Due to the cost benefits of smart charging, these results also indicate that the charging strategy can be a crucial factor for reaching cost parity of PEVs with ICEVs. The study shows that 90 % of Singapore drivers could financially benefit from a PEV in the case of the smart charging strategy as compared to 68 % for uncoordinated charging. These results are, however, based on a hypothetical battery capacity of 10 kWh which was shown to be sufficient for 99 % of Singapore drivers by conducting simulations using CityMoS Traffic. The trend towards larger battery capacities counteracts cost parity with ICEVs. For the uncoordinated charging strategy, the results show that the percentage of Singapore drivers who would financially benefit from a PEV decreases to 40 % for a 20 kWh battery and to only 5 % for a battery capacity of 40 kWh. This shows that mitigating range anxiety by means of larger battery capacities comes at a relatively high cost, especially given the fact that a large share of daily driving needs can be covered with small battery capacities.

The developed framework was also used to investigate the financial viability of energy arbitrage using small-scale BESSs on the example of four different electricity markets. These considerations show that energy arbitrage profits under current conditions are either negative or very moderate which is in accordance with findings in related work. The investigations, however, also identify conditions under which energy arbitrage can be profitable in a sense that it presents a viable business case with a sufficiently high ROI. Apart from a sufficient variability of electricity prices, the appropriate consideration of battery degradation was identified as the most crucial necessary condition to make energy arbitrage financially viable. Under unbeneficial conditions, i.e., battery prices of $ 300 per kWh and access to uniform market prices only, average annual profits range from $ 0 to losses of $ 150 so that none of the investigated markets is profitable for the sole purpose of energy arbitrage. Even at lower battery prices of $ 150 per kWh, only the NEMS is profitable with average annual profits of $ 160. Profits can, however, be considerably increased when relaxing the EOL criterion beyond a remaining capacity of 0.8. The results show that given an EOL capacity of 0.6, annual NEMS profits could be increased from $ 160 to more than $ 400 with similar increases in the other markets. This is not unrealistic for stationary BESSs but also for PEVs it can be argued that the EOL criterion of 0.8 is overly pessimistic.

As can be expected, any kind of electricity price averaging, either temporally or spatially, has an adverse effect on profitability. Profitability can therefore be increased if remuneration is based on NMPs or ZMPs where price fluctuations are higher than for uniform market prices. While the results on the example of zonal prices in the NYISO show that at a battery price of \$ 300 per kWh all zones are still loss-making, moderate profits can be achieved for battery prices of \$ 150 per kWh. Together with the above-mentioned adaptation of the EOL criterion, this can increase profits to a sufficiently high level.

The situation also looks more promising in case energy arbitrage is only supposed to be a secondary purpose of the BESS. In this case, investment costs can be considered as sunk costs so that only operating costs need to be taken into account. In this case, annual profits between \$ 100 and \$ 270 can be achieved with ZMPs and battery prices of \$ 150 per kWh. Further increases of profits up to \$ 618 are observed at a nodal level under these conditions. Also in these cases, profits would further benefit from a less conservative EOL criterion. This is a level where arbitrage profits could make a notable contribution towards the faster amortization of an investment.

The identified conditions for profitable operation may not yet be fulfilled on a large scale. At the current stage, applications would therefore have to be implemented where conditions are most beneficial. Most profitable conditions apply to stationary BESSs serving other primary purposes such as buffering PV generation. These systems could have the benefit of less strict EOL conditions as compared to PEV batteries so that depreciation costs are distributed over a greater lifetime. Furthermore, presuming that the provision of power grid services is not their sole purpose, investment costs do not have to be entirely compensated by arbitrage revenues. Strategically placing these systems at locations where the grid is weak and therefore benefiting from high variability of LMPs can be expected to yield the maximum financial outcome. While these constraints may currently restrict this application to a niche market, decreasing battery prices, increasing battery lifetimes and potentially increasing price variability as a consequence of further RES growth may, however, create the conditions for mass market deployment in the future.

It has to be noted that it is the nature of the considered problem that any quantitative conclusions highly depend upon the chosen parameter values and the investigated technologies. In this regard, the results presented in this work are no exception. The presentation in Chapter 5 with the discussion of quantitative and qualitative influences of the relevant parameters aims to make the established results transferable to a greater context. For accurate assessments, however, individual analyses addressing the particular use case are inevitable. These can be conducted by means of frameworks as the one presented in this work which allow quantitative conclusions for each individual case rather than relying on generalized estimates.

6.2 Outlook

There is a variety of possibilities for further research, development and applications building on the presented work. A brief overview of various potential directions of future work is given in this section.

6.2.1 Extending the Framework's Range of Application

Due to its modular architecture, the developed framework can be extended in different ways to expand its range of application. A number of conceivable extensions are listed as follows:

- **Different ESS types**
  The most obvious extension is to include additional types of storage technologies in order to compare their applicability in different

contexts. These could comprise other BESS technologies such as different types of Li-ion batteries or alternative battery chemistries such as Pb-Ac, Ni-MH or redox-flow batteries. In addition, the application to other types of ESSs can be considered. These could include supercapacitors, thermal ESSs or flywheels, all of which exhibit different characteristics making them applicable for different sorts of applications. Using these different models, the suitability of different kinds of ESSs under certain conditions could be investigated and optimal design parameters be identified.

Different ESS technologies could also be combined to *hybrid energy storage systems* (HESSs) with supplementary operating characteristics in order to better match a particular application scenario. These systems can, for instance, be beneficial if one of the considered applications requires high power capability while another one needs high energy capacity. The framework could be employed for simulation-based sizing of such HESSs and for conducting economic assessments related to their operation.

- **Smart energy hub**

  The framework could also be extended by including an additional generation unit such as a PV panel or a wind turbine. In this case, the scheduler would optimize ESS operation with regard to buffering RES generation and providing power grid services. In addition to electricity price forecasts, RES generation forecasts would additionally be taken into account in order to ensure that the ESS is operated in an optimal range. A further extension could be to include an additional load such as a household, turning the framework into an *energy hub* managing power flows. In a smart home system, this would fulfill the purposes of i) buffering generation of a local RES installation such as a PV module, ii) supplying residential loads during times where demand exceeds RES generation and iii) providing power grid services for peak shaving, valley filling, regulation or reserve.

  Equivalently to the existing implementation, this application would comprise a multistage optimization for charge/discharge scheduling given forecasting information of prices and additionally RES generation and residential load forecasts. Such optimal scheduling will become increasingly relevant as feed-in tariffs for residential PV installations are falling, thus requiring higher shares of self-consumption. Following the premise of minimizing the individual system's costs, the scheduling approach would trade off electricity costs for powering residential loads, revenues from RES generation, battery operating costs and revenues from providing power grid services.

Conceptually similar applications could feature a smart parking lot consisting of PEV charging spots, a RES system and a stationary ESS or a smart bus stop consisting of a RES system, a charging spot for high power supercapacitor charging and an ESS. Possible implementations of the ESS could in these cases, for instance, be batteries or flywheels.

On a technical level, this would require adding modules for RES generation and extending the current storage_commitment module to account for residential loads. Generation and load forecasts may be provided externally or be implemented in the corresponding modules, e.g., by means of time series analysis models. Furthermore, the cost_model would have to be adapted in order to account for the higher dimensionality of the problem.

- **Multi-market participation**
  As outlined in Section 2.3, there are several markets and services which can be considered for BESS participation. Apart from the investigated application for peak shaving and valley filling, reserve and regulation markets are attractive options for BESSs. Depending on the system state, one service may be financially more attractive than the other. In order to provide the service which is of greatest need at a given time and to further increase profitability, the framework could therefore be extended to simultaneously handle multiple markets.

- **Alternative control paradigms**

  A considerably more complex extension would be to implement a transactive control approach, allowing battery agents to negotiate with other entities instead of being treated as price takers. While elements such as multistage optimization and battery degradation monetization implemented already could be preserved in this case, this would require the additional development of iterative negotiation techniques which converge to a decision in a reasonable amount of time.

6.2.2 Development Towards a Scheduling Mechanism Integrated in Battery Management Systems

The presented results show that smarter charging and dispatching mechanisms have the potential for notable financial benefits. Common BESSs, however, operate in a rather simple way by charging at the maximum possible rate provided by the power connection. Charging typically starts once a grid connection is established and it is attempted to reach a full charge. In order to harvest the potential for cost-optimal charging and dispatching, the required technology therefore needs to be implemented in commercially available products. For this purpose, a scheduling device consisting of a combination of hardware and software would have to be integrated into a BESS, e.g., as part of an advanced BMS. In accordance with the presumption that no manual user interaction should be required at any time, this system would have to autonomously manage the charge scheduling considering electricity prices, battery degradation and especially user constraints. The most crucial aspects for realizing such system are listed as follows:

- **Advanced modeling of battery degradation and online monitoring**

  For degradation-aware scheduling, this would require a degradation model to be integrated which appropriately models the degradation behavior of the particular battery type in use. Depending on the way the BMS manages the cell balancing and depending on the degree of heterogeneity of the cells, degradation has to be assessed on the level of individual cells or clusters of similar cells for sufficient accuracy. In the current implementation, the battery pack is assumed to be homogeneous with all cells assuming the same SOC and SOH at the same time. While this assumption is sufficient for the profitability studies under consideration in this work, cells in a real-world system may not be perfectly balanced. This requires an adaptation of the degradation monetization approach to handle a multiplicity of cells at the same time. Through the differentiation between **BatteryPack** and **BatteryCell** objects, the presented framework is conceptualized in a way that allows such differentiation. In this case, a **BatteryPack** has to be composed of different types of **BatteryCell** instances which have to be modeled individually. Due to the large number of cells in a battery pack, this requires further considerations for performance improvements in order to handle the additional computational effort. It also needs to be acknowledged that a model can only approximate the behavior of the real system. With regard to the battery cell, this means that the relevant state variables of the model Φ_{model} and the actual state of the cell Φ_{real} will diverge over time. This may require a continuous re-calibration of the model using online measurement data of an SOH monitoring system.

- **Forecasting of constraints**
  In accordance with the criterion that no manual user interaction should be required, this system would also need to automatically handle constraints resulting from complementary uses of the BESS. In the case of a PEV, this particularly concerns the charging requirements for driving. To ensure the PEV user can complete his trips without running short of energy, a certain battery SOC is required at specific points in time. This needs to be considered by the scheduler as an additional constraint. Users show a great variety of different mobility patterns with some being highly predictable

and some being subject to great temporal variations. This implies that some user groups are more suitable for providing power grid services than others. It is therefore a key success factor to tailor each PEV's charging/discharging strategy to the mobility behavior of its users. This can be achieved by continuously tracking trip data on start and arrival times as well as respective locations as received from a GPS device and by logging energy consumption. Using time series analysis methods, the start time and energy consumption of the next trip can then be predicted. An estimate of the prediction accuracy can further be used to account for sufficient safety buffers. To further improve prediction quality, context data such as the location of the driver collected from smartphone data could be taken into account. In a similar way, constraints could be handled in the case of stationary residential BESSs. As their primary purpose may mostly be the buffering of PV energy, forecasting PV generation and residential load patterns based on historical data as well as weather forecasts could determine the respective constraints.

- **Market interfaces**
  In order to obtain real-time electricity prices and price forecasts, the system requires internet connectivity to connect to the corresponding market. With the ongoing development towards connected cars, this can be expected to be a technical standard in the near future. As different market providers organize their data in different ways, a standardized *application programming interface* (API) is required for obtaining consistent price data. Similarly to mobility pattern prediction, time series analysis, e.g., by means of *autoregressive integrated moving average* (ARIMA) models [17] or other methods related to machine learning [22], could help increase price forecast accuracy. A crucial aspect in this regard is also the development of standardized market participation schemes for distributed BESSs. This may also include the establishment of VPP operators acting as an intermediary between individual

BESSs and the PSO by bundling a large number of batteries to a VPP.

Despite the premise of autonomous operation, the user should always have the possibility of monitoring the system state, specifying safety buffers and degree of risk aversion and, if desired, taking control of the system through mobile communication.

6.2.3 Extended Degradation Models

As any model, the implemented degradation models only provide an approximation of the real-world system. For the presented applications, this is sufficient as the primary question is how a system with the modeled behavior would behave in the considered environment. In a real control setting such as in the system outlined in Section 6.2.2, a closer representation of reality may, however, be required. This could be achieved by the following aspects:

- **Consideration of additional model parameters**
  In order to further refine the degradation cost quantification, a C-rate dependency can be included. As C-rates directly affect temperature because higher currents result in greater energy dissipation, this can be achieved in a first step through implementing a heat generation model which establishes a connection between current and temperature. As temperature is a parameter of the existing model already, this would create a connection between C-rate and degradation.

  The presented studies were performed with a focus on capacity related degradation as the considered applications are expected to be limited by capacity fade. In a different application, the consideration of internal resistance as represented by the parametrization provided in Section 5.2 may, however, also be relevant and could be included accordingly.

- **Battery data crowdsourcing**
  The consideration of battery degradation was presented as an es-

sential part for cost-effective smart charging and profit maximizing provision of power grid services. Correctly quantifying battery degradation costs requires a valid model representing the degradation behavior of the battery used. Since different cell types can show very different aging behaviors, each of them requires its own model. The nature of aging tests, however, is that they are time consuming so that appropriate models may sometimes only become available when the next generation of cells is already on the market. Additionally, not all combinations of operating conditions can be covered so that models may remain incomplete. One way of addressing this issue could be to include advanced diagnostics in commercial BESSs which would keep track of the relevant operating parameters as well as the development of the SOH. The generated data could be transmitted to the battery manufacturer or another entity for analysis. This crowdsourcing of battery data could turn the existing pool of BESSs into a living battery lab. While the testing under uncontrolled conditions would certainly come with its own challenges, the vast amount of data collected for a large variety of operating conditions could possibly improve the understanding of battery aging, serve for the continuous refinement of battery models and speed up battery technology development.

6.2.4 Business Models for Battery Energy Storage Systems

A precondition for the development of business models for BESSs is the existence of market schemes under which distributed BESSs can participate in electricity markets. This requires increased efforts from regulators and market operators to create an environment providing the conditions for viable business cases. In this context, standardized market participation frameworks which are not only applicable to the area covered by an individual market operator would be beneficial. This would require standards for system certification, data exchange and payment so that business developers are not constrained to individual markets. In addition, the development of blockchain-based transaction ledgers can further facilitate the grid integration of distributed BESSs and RES as they allow the implementation of entirely decentralized energy trading mechanisms, thus eliminating the need for central bookkeeping instances.

The results presented in this work indicate that under certain conditions sufficient value can be generated by BESS operation already. The expected decrease of battery prices in conjunction with rising shares of RES will open new windows of opportunity. Opportunities are particularly expected in the domain of small-scale home batteries for the purpose of buffering energy generated by residential PV systems. The purchase of these residential BESSs is partly stimulated by the trend of changing feed-in laws which require owners of PV systems to increase self-consumption as well as by government subsidies supporting the purchase of home batteries. As compared to PEV batteries, stationary batteries benefit from less strictly defined EOL conditions, thus lowering operating costs compared to PEVs. Furthermore, their operation is less constrained by behavioral aspects such as range anxiety. These aspects make them particularly suitable for different types of DSM services. At the same time, there is a lack of simple financing, ownership and remuneration models which would make it easy enough for potential customers to acquire and operate systems for these purposes. This requires plug-and-play solutions which integrate the entire value chain from financing to optimal system operation. By pooling the financial risk of operating BESSs through appropriate financing and remuneration concepts, costs and benefits of BESS operation could become more transparent, thus extending consumer interest beyond the group of early adopters. Frameworks as developed in this work can contribute to assessing the financial viability of different business models and help to identify and prioritize those customers for whom BESS installations are most profitable.

Bibliography

[1] J. Bishop, C. Axon, D. Bonilla, M. Tran, D. Banister, and M. McCulloch. Evaluating the impact of V2G services on the degradation of batteries in PHEV and EV. *Applied Energy*, 111: 206–218, 2013. doi:10.1016/j.apenergy.2013.04.094.

[2] J. Bishop, C. Axon, D. Bonilla, and D. Banister. Estimating the grid payments necessary to compensate additional costs to prospective electric vehicle owners who provide vehicle-to-grid ancillary services. *Energy*, 94:715–727, 2016. doi:10.1016/j.energy.2015.11.029.

[3] S.-L. Andersson, A. Elofsson, M. Galus, L. Göransson, S. Karlsson, F. Johnsson, and G. Andersson. Plug-in hybrid electric vehicles as regulating power providers: Case studies of Sweden and Germany. *Energy Policy*, 38(6):2751–2762, 2010. ISSN 03014215. doi:10.1016/j.enpol.2010.01.006.

[4] O. Egbue and S. Long. Barriers to widespread adoption of electric vehicles: An analysis of consumer attitudes and perceptions. *Energy Policy*, 48:717–729, 2012. doi:10.1016/j.enpol.2012.06.009.

[5] D. Ciechanowicz, D. Pelzer, and A. Knoll. Simulation-based approach for investigating the impact of electric vehicles on power grids. In *IEEE PES Asia-Pacific Power and Energy Engineering Conference (APPEEC)*, 2015. doi:10.1109/APPEEC.2015.7381046.

[6] D. Ciechanowicz, D. Pelzer, B. Bartenschlager, and A. Knoll. A modular power system planning and power flow simulation framework for generating and evaluating power network models. *IEEE Transactions on Power Systems*, 32(3):2214–2224, 2017. ISSN 0885-8950. doi:10.1109/TPWRS.2016.2602479.

© Springer Fachmedien Wiesbaden GmbH, part of Springer Nature 2019
D. Pelzer, *A Modular Framework for Optimizing Grid Integration of Mobile and Stationary Energy Storage in Smart Grids*, https://doi.org/10.1007/978-3-658-27024-7

[7] D. Ciechanowicz. *A power system planning and power flow simulation framework for generating and evaluating power network models.* PhD thesis, Technical University of Munich, 2018.

[8] V. Viswanathan, D. Zehe, J. Ivanchev, D. Pelzer, A. Knoll, and H. Aydt. Simulation-assisted exploration of charging infrastructure requirements for electric vehicles in urban environments. *Journal of Computational Science*, 12:1 – 10, 2016. ISSN 1877-7503. doi:10.1016/j.jocs.2015.10.012.

[9] R. Bi, J. Xiao, D. Pelzer, D. Ciechanowicz, D. Eckhoff, and A. Knoll. A simulation-based heuristic for city-scale electric vehicle charging station placement. In *IEEE International Conference on Intelligent Transportation Systems (ITSC)*, 2017.

[10] D. Pelzer, J. Xiao, D. Zehe, M. Lees, A. Knoll, and H. Aydt. A partition-based match making algorithm for dynamic ridesharing. *IEEE Transactions on Intelligent Transportation Systems*, 16(5):2587–2598, 2015. ISSN 1524-9050. doi:10.1109/TITS.2015.2413453.

[11] R. Bessa and M. Matos. Economic and technical management of an aggregation agent for electric vehicles: A literature survey. *European Transactions on Electrical Power*, 22(3):334–350, 2012. doi:10.1002/etep.565.

[12] C. Guille and G. Gross. A conceptual framework for the vehicle-to-grid (V2G) implementation. *Energy Policy*, 37(11):4379–4390, 2009. doi:10.1016/j.enpol.2009.05.053.

[13] A. Raab, M. Ferdowsi, E. Karfopoulos, I. Unda, S. Skarvelis-Kazakos, P. Papadopoulos, E. Abbasi, L. Cipcigan, N. Jenkins, N. Hatziargyriou, and K. Strunz. Virtual power plant control concepts with electric vehicles. In *16th International Conference on Intelligent System Applications to Power Systems*, 2011. doi:10.1109/ISAP.2011.6082214.

[14] R. Bessa, M. Matos, F. Soares, and J. Lopes. Optimized bidding of a EV aggregation agent in the electricity market. *IEEE Transactions on Smart Grid*, 3(1):443–452, 2012. doi:10.1109/TSG.2011.2159632.

[15] R. Bessa and M. Matos. Optimization models for EV aggregator participation in a manual reserve market. *IEEE Transactions on Power Systems*, 28(3):3085–3095, 2013. doi:10.1109/TPWRS.2012.2233222.

[16] S. Han, S. Han, and K. Sezaki. Development of an optimal vehicle-to-grid aggregator for frequency regulation. *IEEE Transactions on Smart Grid*, 1(1):65–72, 2010. doi:10.1109/TSG.2010.2045163.

[17] I. Momber, A. Siddiqui, T. Roman, and L. Soder. Risk averse scheduling by a PEV aggregator under uncertainty. *IEEE Transactions on Power Systems*, 30(2):882–891, 2015. doi:10.1109/TPWRS.2014.2330375.

[18] B. Jansen, C. Binding, O. Sundstrom, and D. Gantenbein. Architecture and communication of an electric vehicle virtual power plant. In *1st IEEE International Conference on Smart Grid Communications (SmartGridComm)*, pages 149–154, 2010. doi:10.1109/SMARTGRID.2010.5622033.

[19] P. Palensky and D. Dietrich. Demand side management: Demand response, intelligent energy systems, and smart loads. *IEEE Transactions on Industrial Informatics*, 7(3):381–388, 2011. doi:10.1109/TII.2011.2158841.

[20] M. Rostami and M. Raoofat. Optimal operating strategy of virtual power plant considering plug-in hybrid electric vehicles load. *International Transactions on Electrical Energy Systems*, 2015. doi:10.1002/etep.2074.

[21] G. Chalkiadakis, V. Robu, R. Kota, A. Rogers, and N. Jennings. Cooperatives of distributed energy resources for efficient virtual

power plants. In *10th International Conference on Autonomous Agents and Multiagent Systems*, volume 2, pages 737–744. International Foundation for Autonomous Agents and Multiagent Systems, 2011. ISBN 0-9826571-6-1, 978-0-9826571-6-4.

[22] S. Ramchurn, P. Vytelingum, A. Rogers, and N. Jennings. Putting the 'smarts' into the smart grid: A grand challenge for artificial intelligence. *Communications of the ACM*, 55(4): 86–97, 2012. ISSN 0001-0782. doi:10.1145/2133806.2133825.

[23] S. Kamboj, W. Kempton, and K. Decker. Deploying power grid-integrated electric vehicles as a multi-agent system. In *10th International Conference on Autonomous Agents and Multiagent Systems*, volume 1, pages 9–16. International Foundation for Autonomous Agents and Multiagent Systems, 2011. ISBN 0-9826571-5-3, 978-0-9826571-5-7.

[24] A. L. Dimeas and N. D. Hatziargyriou. Agent based control of virtual power plants. In *International Conference on Intelligent Systems Applications to Power Systems*, pages 1–6, 2007. doi:10.1109/ISAP.2007.4441671.

[25] F. Sioshansi. So what's so smart about the smart grid? *Electricity Journal*, 24(10):91–99, 2011. doi:10.1016/j.tej.2011.11.005.

[26] G. Verbong, S. Beemsterboer, and F. Sengers. Smart grids or smart users? Involving users in developing a low carbon electricity economy. *Energy Policy*, 52:117–125, 2013. doi:10.1016/j.enpol.2012.05.003.

[27] Y. Parag and B. K. Sovacool. Electricity market design for the prosumer era. *Nature Energy*, 1, 2016. doi:doi:10.1038/nenergy.2016.32.

[28] B. M. Buchholz and Z. A. Styczynski. *Smart Grids - fundamentals and technologies in electricity networks*. Springer, Heidelberg, 2014. ISBN 3642451209. doi:10.1007/978-3-642-45120-1.

[29] J. Momoh. *Smart grid: Fundamentals of design and analysis.* Wiley-IEEE Press, 2012. ISBN 978-0-470-88939-8.

[30] S. Rajakaruna, A. Garcia-Cerrada, and A. Ghosh. *Plug-in electric vehicles in smart grids: Integration techniques.* Springer, Singapore, 2015. ISBN 978-981-287-298-2.

[31] S. Rajakaruna, F. Shahnia, and A. Ghosh. *Plug-in electric vehicles in smart grids: Energy management.* Springer, Singapore, 2015. ISBN 978-981-287-301-9.

[32] B. Dunn, H. Kamath, and J.-M. Tarascon. Electrical energy storage for the grid: A battery of choices. *Science,* 334(6058): 928–935, 2011. doi:10.1126/science.1212741.

[33] L. Gelazanskas and K. Gamage. Demand side management in smart grid: A review and proposals for future direction. *Sustainable Cities and Society,* 11:22–30, 2014. doi:10.1016/j.scs.2013.11.001.

[34] P. Siano. Demand response and smart grids - a survey. *Renewable and Sustainable Energy Reviews,* 30:461 – 478, 2014. ISSN 1364-0321. doi:10.1016/j.rser.2013.10.022.

[35] M. Albadi and E. El-Saadany. A summary of demand response in electricity markets. *Electric Power Systems Research,* 78(11): 1989–1996, 2008. doi:10.1016/j.epsr.2008.04.002.

[36] C. Gellings. The concept of demand-side management for electric utilities. *Proceedings of the IEEE,* 73(10):1468–1470, 1985. doi:10.1109/PROC.1985.13318.

[37] T. Li and M. Shahidehpour. Price-based unit commitment: A case of Lagrangian relaxation versus mixed integer programming. *IEEE Transactions on Power Systems,* 20(4):2015–2025, 2005. ISSN 0885-8950. doi:10.1109/TPWRS.2005.857391.

[38] J. García-Villalobos, I. Zamora, J. San Martín, F. Asensio, and V. Aperribay. Plug-in electric vehicles in electric dis-

tribution networks: A review of smart charging approaches. *Renewable and Sustainable Energy Reviews*, 38:717–731, 2014. doi:10.1016/j.rser.2014.07.040.

[39] D. Dallinger and M. Wietschel. Grid integration of intermittent renewable energy sources using price-responsive plug-in electric vehicles. *Renewable and Sustainable Energy Reviews*, 16(5):3370–3382, 2012. doi:10.1016/j.rser.2012.02.019.

[40] AEP Ohio. gridSmart demonstration project: Final technical report. Technical report, AEP Ohio, 2014.

[41] R. Melton. Annual report: Pacific Northwest smart grid demonstration project. Technical report, 2013.

[42] B. van der Waaij, W. Wijbrandi, and M. Konsman. White paper energy flexibility platform and interface (EF-Pi). Technical report, TNO, 2015.

[43] T. W. Malone. Modeling coordination in organizations and markets. *Management Science*, 33(10):1317–1332, 1987. doi:10.1287/mnsc.33.10.1317.

[44] M. Galus, R. Waraich, and G. Andersson. Predictive, distributed, hierarchical charging control of PHEVs in the distribution system of a large urban area incorporating a multi agent transportation simulation. Technical report, ETH Zürich, 2011.

[45] M. Caramanis and J. Foster. Management of electric vehicle charging to mitigate renewable generation intermittency and distribution network congestion. In *Proceedings of the 48th IEEE Conference on Decision and Control (CDC)*, pages 4717–4722, 2009. doi:10.1109/CDC.2009.5399955.

[46] Q. Wang, C. Zhang, Y. Ding, G. Xydis, J. Wang, and J. Østergaard. Review of real-time electricity markets for integrating distributed energy resources and demand response. *Applied Energy*, 138:695 – 706, 2015. ISSN 0306-2619. doi:10.1016/j.apenergy.2014.10.048.

Bibliography 183

[47] A. Srivastava, S. Kamalasadan, D. Patel, S. Sankar, and K. Al-Olimat. Electricity markets: An overview and comparative study. *International Journal of Energy Sector Management*, 5 (2):169–200, 2011. doi:10.1108/17506221111145977.

[48] J. Hu, H. Morais, T. Sousa, and M. Lind. Electric vehicle fleet management in smart grids: A review of services, optimization and control aspects. *Renewable and Sustainable Energy Reviews*, 56:1207–1226, 2016. doi:10.1016/j.rser.2015.12.014.

[49] W. Kempton and J. Tomić. Vehicle-to-grid power fundamentals: Calculating capacity and net revenue. *Journal of Power Sources*, 144(1):268–279, 2005. ISSN 03787753. doi:10.1016/j.jpowsour.2004.12.025.

[50] M. Palacín and A. de Guibert. Batteries: Why do batteries fail? *Science*, 351(6273), 2016. doi:10.1126/science.1253292.

[51] J. Goodenough and Y. Kim. Challenges for rechargeable Li batteries. *Chemistry of Materials*, 22(3):587–603, 2010. doi:10.1021/cm901452z.

[52] J.-M. Tarascon and M. Armand. Issues and challenges facing rechargeable lithium batteries. *Nature*, 414(6861):359–367, 2001. doi:10.1038/35104644.

[53] K. Clement-Nyns. *Impact of plug-in hybrid electric vehicles on the electricity system*. PhD thesis, University of Leuven, 2010.

[54] D. Linden and T. B. Reddy. *Linden's handbook of batteries*. McGraw-Hill, 2001. ISBN 978-0071624213.

[55] Y. Nishi. Lithium ion secondary batteries; past 10 years and the future. *Journal of Power Sources*, 100(1):101 – 106, 2001. ISSN 0378-7753. doi:10.1016/S0378-7753(01)00887-4.

[56] A. Dinger, M. Ripley, X. Mosquet, M. Rabl, D. Rizoulis, M. Russo, and G. Sticher. Batteries for electric cars - chal-

lenges opportunities and the outlook to 2020. Technical report, The Boston Consulting Group, 2010.

[57] B. Nykvist and M. Nilsson. Rapidly falling costs of battery packs for electric vehicles. *Nature Climate Change*, 5(4):329–332, 2015. doi:10.1038/nclimate2564.

[58] S. Knupfer, R. Hensley, P. Hertzke, P. Schaufuss, N. Laverty, and N. Kramer. Electrifying insights: How automakers can drive electrified vehicle sales and profitability. Technical report, McKinsey & Company, 2017.

[59] N. Kittner, F. Lill, and D. M. Kammen. Energy storage deployment and innovation for the clean energy transition. *Nature Energy*, 2:17125, 2017. doi:10.1038/nenergy.2017.125.

[60] Tesla Motors. Tesla gigafactory. URL https://www.tesla. com/en_GB/gigafactory.

[61] Federal Energy Regulatory Commission (FERC). Electric storage participation in markets operated by regional transmission organizations and independent system operators. Technical report, 2016.

[62] Deutsche Übertragungsnetzbetreiber. Eckpunkte und Freiheitsgrade bei Erbringung von Primärregelleistung. Technical report, 2014.

[63] Deutsche Übertragungsnetzbetreiber. Anforderungen an die Speicherkapazität bei Batterien für die Primärregelleistung. Technical report, 2015.

[64] S. Ghanbari. Development of a business model for intelligent control of battery storage in the german reserve market. Master's thesis, Technical University of Munich, 2015.

[65] International Energy Agency (IEA). Global EV outlook 2017. Technical report, 2017.

[66] R. Hensley, J. Newman, and M. Rogers. Battery technology charges ahead. Technical report, McKinsey & Company, 2012.

[67] Deutsche Bank. The end of the oil age. 2011 and beyond: A reality check. Technical report, 2000.

[68] M. Hackmann, H. Pyschny, and R. Stanek. Total Cost of Ownership Analyse für Elektrofahrzeuge. Technical report, 2014.

[69] S. Hadley. Evaluating the impact of plug-in hybrid electric vehicles on regional electricity supplies. In *iREP Symposium - Bulk Power System Dynamics and Control - VII. Revitalizing Operational Reliability*, 2007. doi:10.1109/IREP.2007.4410538.

[70] M. Kintner-Meyer, K. Schneider, and R. Pratt. Impacts assessment of plug-in hybrid vehicles on electric utilities and regional US power grids, Part 1: Technical analysis. Technical report, Pacific Northwest National Laboratory, 2007.

[71] Z. Darabi and M. Ferdowsi. Examining power grid's capacity to meet transportation electrification demand. In *IEEE Power and Energy Society General Meeting*, 2012. doi:10.1109/PESGM.2012.6345204.

[72] H. Gerbracht, D. Möst, and W. Fichtner. Impacts of plug-in electric vehicles on Germany's power plant portfolio - a model based approach. In *7th International Conference on the European Energy Market*, 2010. doi:10.1109/EEM.2010.5558696.

[73] W.-J. Park, K.-B. Song, and J.-W. Park. Impact of electric vehicle penetration-based charging demand on load profile. *Journal of Electrical Engineering and Technology*, 8(2):244–251, 2013. doi:10.5370/JEET.2013.8.2.244.

[74] M. Galus, M. Vayá, T. Krause, and G. Andersson. The role of electric vehicles in smart grids. *Wiley Interdisciplinary Reviews: Energy and Environment*, 2(4):384–400, 2013. doi:10.1002/wene.56.

[75] N. Hartmann and E. Özdemir. Impact of different utilization scenarios of electric vehicles on the German grid in 2030. *Journal of Power Sources*, 196(4):2311–2318, 2011. doi:10.1016/j.jpowsour.2010.09.117.

[76] R. Green, L. Wang, and M. Alam. The impact of plug-in hybrid electric vehicles on distribution networks: A review and outlook. *Renewable and Sustainable Energy Reviews*, 15(1):544–553, 2011. doi:10.1016/j.rser.2010.08.015.

[77] K. Clement-Nyns, E. Haesen, and J. Driesen. The impact of vehicle-to-grid on the distribution grid. *Electric Power Systems Research*, 81(1):185–192, 2011. ISSN 03787796. doi:10.1016/j.epsr.2010.08.007.

[78] G. Razeghi, L. Zhang, T. Brown, and S. Samuelsen. Impacts of plug-in hybrid electric vehicles on a residential transformer using stochastic and empirical analysis. *Journal of Power Sources*, 252:277–285, 2014. doi:10.1016/j.jpowsour.2013.11.089.

[79] E. Akhavan-Rezai, M. Shaaban, E. El-Saadany, and A. Zidan. Uncoordinated charging impacts of electric vehicles on electric distribution grids: Normal and fast charging comparison. In *IEEE Power and Energy Society General Meeting*, 2012. doi:10.1109/PESGM.2012.6345583.

[80] L. Dow, M. Marshall, L. Xu, J. Agüero, and H. Willis. A novel approach for evaluating the impact of electric vehicles on the power distribution system. In *IEEE PES General Meeting*, 2010. doi:10.1109/PES.2010.5589507.

[81] J. A. P. Lopes, F. J. Soares, and P. M. R. Almeida. Identifying management procedures to deal with connection of electric vehicles in the grid. In *IEEE Bucharest PowerTech*, pages 1–8, 2009. doi:10.1109/PTC.2009.5282155.

[82] A. Aljanad and A. Mohamed. Impact of plug-in hybrid electric vehicle on power distribution system considering vehicle

to grid technology: A review. *Research Journal of Applied Sciences, Engineering and Technology*, 10(12):1404–1413, 2015. ISSN 2040 7459.

[83] S. Habib, M. Kamran, and U. Rashid. Impact analysis of vehicle-to-grid technology and charging strategies of electric vehicles on distribution networks - a review. *Journal of Power Sources*, 277:205–214, 2015. doi:10.1016/j.jpowsour.2014.12.020.

[84] K. Tan, V. Ramachandaramurthy, and J. Yong. Integration of electric vehicles in smart grid: A review on vehicle to grid technologies and optimization techniques. *Renewable and Sustainable Energy Reviews*, 53:720–732, 2016. doi:10.1016/j.rser.2015.09.012.

[85] M. Yilmaz and P. Krein. Review of the impact of vehicle-to-grid technologies on distribution systems and utility interfaces. *IEEE Transactions on Power Electronics*, 28(12):5673–5689, 2013. doi:10.1109/TPEL.2012.2227500.

[86] A. Rogers, S. Ramchurn, and N. Jennings. Delivering the smart grid: Challenges for autonomous agents and multi-agent systems research. In *Proceedings of the 26th AAAI Conference on Artificial Intelligence*, volume 3, pages 2166–2172, 2012. ISBN 978-157735568-7.

[87] C. Flath, J. Ilg, S. Gottwalt, H. Schmeck, and C. Weinhardt. Improving electric vehicle charging coordination through area pricing. *Transportation Science*, 48(4):619–634, 2014. doi:10.1287/trsc.2013.0467.

[88] Z. Fan. A distributed demand response algorithm and its application to PHEV charging in smart grids. *IEEE Transactions on Smart Grid*, 3(3):1280–1290, 2012. doi:10.1109/TSG.2012.2185075.

[89] H. Seifi and M. S. Sepasian. *Electric power system planning: Issues, algorithms and solutions.* Springer, Berlin, Heidelberg, 2011. doi:10.1007/978-3-642-17989-1.

[90] Streetdirectory. SG & Singapore map. URL http://www.streetdirectory.com.

[91] SLA. OneMap. URL http://www.onemap.sg.

[92] Energy Market Authority (EMA) of Singapore. Monthly electricity consumption by sector and half-hourly system demand data, 2015. URL http://www.ema.gov.sg/Statistics.aspx.

[93] Energy Market Authority (EMA) of Singapore. Licensed generation capacity by generation company, 2015. URL http://www.ema.gov.sg/Statistics.aspx.

[94] Nexans. Power cables 1-30 kV and high voltage cables for power transmission. URL http://www.nexans.de.

[95] J. C. Mukherjee and A. Gupta. A review of charge scheduling of electric vehicles in smart grid. *IEEE Systems Journal*, 9(4):1541–1553, 2015. ISSN 1932-8184. doi:10.1109/JSYST.2014.2356559.

[96] A. Hota, M. Juvvanapudi, and P. Bajpai. Issues and solution approaches in PHEV integration to the smart grid. *Renewable and Sustainable Energy Reviews*, 30:217–229, 2014. doi:10.1016/j.rser.2013.10.008.

[97] A. Schuller. Charging coordination paradigms of electric vehicles. *Power Systems*, 88, 2015. doi:10.1007/978-981-287-317-0_1.

[98] F. Mwasilu, J. Justo, E.-K. Kim, T. Do, and J.-W. Jung. Electric vehicles and smart grid interaction: A review on vehicle to grid and renewable energy sources integration. *Renewable and Sustainable Energy Reviews*, 34:501–516, 2014. doi:10.1016/j.rser.2014.03.031.

[99] X. Zhang, Q. Wang, G. Xu, and Z. Wu. A review of plug-in electric vehicles as distributed energy storages in smart grid. In *IEEE PES Innovative Smart Grid Technologies Conference, Europe (ISGT)*, volume 2015-January, 2015. doi:10.1109/ISGTEurope.2014.7028853.

[100] D. Richardson. Electric vehicles and the electric grid: A review of modeling approaches, impacts, and renewable energy integration. *Renewable and Sustainable Energy Reviews*, 19:247–254, 2013. doi:10.1016/j.rser.2012.11.042.

[101] W. Kempton and S. E. Letendre. Electric vehicles as a new power source for electric utilities. *Transportation Research Part D: Transport and Environment*, 2(3):157–175, 1997. doi:10.1016/S1361-9209(97)00001-1.

[102] L. Pieltain Fernández, T. Gómez San Román, R. Cossent, C. Mateo Domingo, and P. Frías. Assessment of the impact of plug-in electric vehicles on distribution networks. *IEEE Transactions on Power Systems*, 26(1):206–213, 2011. doi:10.1109/TPWRS.2010.2049133.

[103] E. Sortomme and M. El-Sharkawi. Optimal charging strategies for unidirectional vehicle-to-grid. *IEEE Transactions on Smart Grid*, 2(1):119–126, 2011. doi:10.1109/TSG.2010.2090910.

[104] E. Sortomme and M. El-Sharkawi. Optimal scheduling of vehicle-to-grid energy and ancillary services. *IEEE Transactions on Smart Grid*, 3(1):351–359, 2012. doi:10.1109/TSG.2011.2164099.

[105] J. Tomić and W. Kempton. Using fleets of electric-drive vehicles for grid support. *Journal of Power Sources*, 168(2):459–468, 2007. ISSN 03787753. doi:10.1016/j.jpowsour.2007.03.010.

[106] S. B. Peterson, J. F. Whitacre, and J. Apt. The economics of using plug-in hybrid electric vehicle battery packs for grid stor-

age. *Journal of Power Sources*, 195(8):2377–2384, 2010. ISSN 03787753. doi:10.1016/j.jpowsour.2009.09.070.

[107] A. Bhatti, Z. Salam, M. Aziz, K. Yee, and R. Ashique. Electric vehicles charging using photovoltaic: Status and technological review. *Renewable and Sustainable Energy Reviews*, 54:34–47, 2016. doi:10.1016/j.rser.2015.09.091.

[108] A. Schuller, B. Dietz, C. Flath, and C. Weinhardt. Charging strategies for battery electric vehicles: Economic benchmark and V2G potential. *IEEE Transactions on Power Systems*, 29 (5):2014–2222, 2014. doi:10.1109/TPWRS.2014.2301024.

[109] D. Ciechanowicz, A. Knoll, P. Osswald, and D. Pelzer. Towards a business case for vehicle-to-grid–maximizing profits in ancillary service markets. In S. Rajakaruna, F. Shahnia, and A. Ghosh, editors, *Plug-in electric vehicles in smart grids*. Springer, Singapore, 2015. ISBN 978-981-287-301-9. doi:10.1007/978-981-287-302-6_8.

[110] D. Pelzer, D. Ciechanowicz, H. Aydt, and A. Knoll. A price-responsive dispatching strategy for vehicle-to-grid: An economic evaluation applied to the case of Singapore. *Journal of Power Sources*, 256(0):345 – 353, 2014. ISSN 0378-7753. doi:10.1016/j.jpowsour.2014.01.076.

[111] B. Lunz, Z. Yan, J. Gerschler, and D. Sauer. Influence of plug-in hybrid electric vehicle charging strategies on charging and battery degradation costs. *Energy Policy*, 46:511–519, 2012. doi:10.1016/j.enpol.2012.04.017.

[112] C. Guenther, B. Schott, W. Hennings, P. Waldowski, and M. Danzer. Model-based investigation of electric vehicle battery aging by means of vehicle-to-grid scenario simulations. *Journal of Power Sources*, 239:604–610, 2013. doi:10.1016/j.jpowsour.2013.02.041.

[113] R. Dufo-López. Optimisation of size and control of grid-connected storage under real time electricity pricing conditions. *Applied Energy*, 140:395–408, 2015. doi:10.1016/j.apenergy.2014.12.012.

[114] A. Shcherbakova, A. Kleit, and J. Cho. The value of energy storage in South Korea's electricity market: A hotelling approach. *Applied Energy*, 125:93–102, 2014. doi:10.1016/j.apenergy.2014.03.046.

[115] C. Zhou, K. Qian, M. Allan, and W. Zhou. Modeling of the cost of EV battery wear due to V2G application in power systems. *IEEE Transactions on Energy Conversion*, 26(4):1041–1050, 2011. doi:10.1109/TEC.2011.2159977.

[116] R. Sioshansi, P. Denholm, T. Jenkin, and J. Weiss. Estimating the value of electricity storage in PJM: Arbitrage and some welfare effects. *Energy Economics*, 31(2):269–277, 2009. doi:10.1016/j.eneco.2008.10.005.

[117] T. Kristoffersen, K. Capion, and P. Meibom. Optimal charging of electric drive vehicles in a market environment. *Applied Energy*, 88(5):1940–1948, 2011. doi:10.1016/j.apenergy.2010.12.015.

[118] M. González Vayá, T. Krause, R. Waraich, and G. Andersson. Locational marginal pricing based impact assessment of plug-in hybrid electric vehicles on transmission networks. In *Cigre 2011 Bologna Symposium*, 2011. ISBN 978-285873165-7.

[119] L. Agarwal, W. Peng, and L. Goel. Using EV battery packs for vehicle-to-grid applications: An economic analysis. In *IEEE Innovative Smart Grid Technologies, Asia (ISGT)*, pages 663–668, 2014. doi:10.1109/ISGT-Asia.2014.6873871.

[120] D. Dallinger, D. Krampe, and M. Wietschel. Vehicle-to-grid regulation reserves based on a dynamic simulation of mobil-

ity behavior. *IEEE Transactions on Smart Grid*, 2(2):302–313, 2011. ISSN 1949-3053. doi:10.1109/TSG.2011.2131692.

[121] N. Rotering and M. Ilic. Optimal charge control of plug-in hybrid electric vehicles in deregulated electricity markets. *IEEE Transactions on Power Systems*, 26(3):1021–1029, 2011. doi:10.1109/TPWRS.2010.2086083.

[122] E. L. Karfopoulos and N. D. Hatziargyriou. Distributed coordination of electric vehicles providing V2G services. *IEEE Transactions on Power Systems*, 31(1):329–338, 2016. ISSN 0885-8950. doi:10.1109/TPWRS.2015.2395723.

[123] E. Iversen, J. Morales, and H. Madsen. Optimal charging of an electric vehicle using a markov decision process. *Applied Energy*, 123:1–12, 2014. doi:10.1016/j.apenergy.2014.02.003.

[124] A. Hoke, A. Brissette, D. Maksimovic, D. Kelly, A. Pratt, and D. Boundy. Maximizing lithium ion vehicle battery life through optimized partial charging. In *IEEE PES Innovative Smart Grid Technologies (ISGT)*, pages 1–5, 2013. doi:10.1109/ISGT.2013.6497818.

[125] Z. Ma, S. Zou, and X. Liu. A distributed charging coordination for large-scale plug-in electric vehicles considering battery degradation cost. *IEEE Transactions on Control Systems Technology*, 23(5):2044–2052, 2015. ISSN 1063-6536. doi:10.1109/TCST.2015.2394319.

[126] A. E. Trippe, R. Arunachala, T. Massier, A. Jossen, and T. Hamacher. Charging optimization of battery electric vehicles including cycle battery aging. In *IEEE PES Innovative Smart Grid Technologies, Europe (ISGT)*, pages 1–6, 2014. doi:10.1109/ISGTEurope.2014.7028735.

[127] M. Honarmand, A. Zakariazadeh, and S. Jadid. Integrated scheduling of renewable generation and electric vehicles parking

lot in a smart microgrid. *Energy Conversion and Management*, 86:745–755, 2014. doi:10.1016/j.enconman.2014.06.044.

[128] J. Tan and L. Wang. A game-theoretic framework for vehicle-to-grid frequency regulation considering smart charging mechanism. *IEEE Transactions on Smart Grid*, 2016. doi:10.1109/TSG.2016.2524020.

[129] T. A. Nguyen and M. L. Crow. Stochastic optimization of renewable-based microgrid operation incorporating battery operating cost. *IEEE Transactions on Power Systems*, 31(3):2289–2296, 2016. ISSN 0885-8950. doi:10.1109/TPWRS.2015.2455491.

[130] G. He, Q. Chen, C. Kang, P. Pinson, and Q. Xia. Optimal bidding strategy of battery storage in power markets considering performance-based regulation and battery cycle life. *IEEE Transactions on Smart Grid*, 7(5):2359–2367, 2016. ISSN 1949-3053. doi:10.1109/TSG.2015.2424314.

[131] A. Hoke, A. Brissette, K. Smith, A. Pratt, and D. Maksimovic. Accounting for lithium-ion battery degradation in electric vehicle charging optimization. *IEEE Journal of Emerging and Selected Topics in Power Electronics*, 2(3):691–700, 2014. ISSN 2168-6777. doi:10.1109/JESTPE.2014.2315961.

[132] A. Marongiu, M. Roscher, and D. Sauer. Influence of the vehicle-to-grid strategy on the aging behavior of lithium battery electric vehicles. *Applied Energy*, 137:899–912, 2015. doi:10.1016/j.apenergy.2014.06.063.

[133] S. Han, S. Han, and H. Aki. A practical battery wear model for electric vehicle charging applications. *Applied Energy*, 113: 1100–1108, 2014. doi:10.1016/j.apenergy.2013.08.062.

[134] M. Ouyang, X. Feng, X. Han, L. Lu, Z. Li, and X. He. A dynamic capacity degradation model and its applications con-

sidering varying load for a large format Li-ion battery. *Applied Energy*, 165:48–59, 2016. doi:10.1016/j.apenergy.2015.12.063.

[135] E. Camacho and C. Bordons Alba. *Model predictive control.* Springer, London, 2007. ISBN 978-1-85233-694-3.

[136] T. Bray, J. Paoli, C. Sperberg-McQueen, E. Maler, and F. Yergeau. W3C Extensible Markup Language (XML) 1.0. *W3C Recommendation*, 2008.

[137] S. Gao, C. Sperberg-McQueen, and H. Thompson. W3C XML Schema Definition Language (XSD) 1.1 Part 1: Structures. *W3C Recommendation*, 2012.

[138] D. Peterson, S. Gao, A. Malhotra, C. Sperberg-McQueen, and H. Thompson. W3C XML Schema Definition Language (XSD) 1.1 Part 2: Datatypes. *W3C Recommendation*, 2012.

[139] Y. Xu, H. Aydt, and M. Lees. SEMSim: A distributed architecture for multi-scale traffic simulation. In *ACM/IEEE/SCS 26th Workshop on Principles of Advanced and Distributed Simulation (PADS)*, pages 178–180, 2012. doi:10.1109/PADS.2012.40.

[140] Navteq. We are HERE. URL http://navteq.com.

[141] H. Ortega-Arranz, D. Llanos, and A. Gonzalez-Escribano. The shortest-path problem: Analysis and comparison of methods. *Synthesis Lectures on Theoretical Computer Science*, 1(1):1–87, 2014. doi:10.2200/s00618ed1v01y201412tcs001.

[142] E. Bonabeau. Agent-based modeling: Methods and techniques for simulating human systems. *Proceedings of the National Academy of Sciences of the United States of America*, 99 (SUPPL. 3):7280–7287, 2002. doi:10.1073/pnas.082080899.

[143] M. Treiber, A. Hennecke, and D. Helbing. Congested traffic states in empirical observations and microscopic simulations. *Physical Review E - Statistical Physics, Plasmas, Fluids,*

and *Related Interdisciplinary Topics*, 62(2 B):1805–1824, 2000. doi:10.1103/PhysRevE.62.1805.

[144] P. Gipps. A behavioural car-following model for computer simulation. *Transportation Research Part B*, 15(2):105–111, 1981. doi:10.1016/0191-2615(81)90037-0.

[145] R. Bi, J. Xiao, V. Viswanathan, and A. Knoll. Influence of charging behaviour given charging infrastructure specification: A case study of Singapore. *Journal of Computational Science*, 20:118–128, 2017. doi:10.1016/j.jocs.2017.03.013.

[146] Land Transport Authority (LTA). Household interview travel survey, 2012. URL http://www.lta.gov.sg/apps/news/page. aspx?c=2&id=1b6b1e1e-f727-43bb-8688-f589056ad1c4.

[147] M. Ackerman, S. Ben-David, S. Brânzei, and D. Loker. Weighted clustering. *arXiv preprint arXiv:1109.1844*, 2011.

[148] M. Ester, H. P. Kriegel, J. Sander, and X. Xu. A density-based algorithm for discovering clusters in large spatial databases with noise. In E. Simoudis, J. Han, and U. Fayyad, editors, *Second International Conference on Knowledge Discovery and Data Mining*, pages 226–231, Portland, Oregon, 1996. AAAI Press.

[149] G. A. Croes. A method for solving Traveling-Salesman Problems. *Operations Research*, 6:791–812, 1958. doi:10.1287/opre.6.6.791.

[150] J. B. Kruskal. On the shortest spanning subtree of a graph and the Traveling Salesman Problem. *Proceedings of the American Mathematical Society*, 7(1):48–50, 1956. doi:10.2307/2033241.

[151] L. Guibas, D. E. Knuth, and M. Sharir. Randomized incremental construction of Delaunay and Voronoi diagrams. *Algorithmica*, 7(1-6):381–413, 1992. ISSN 0178-4617. doi:10.1007/BF01758770.

[152] R. Lincoln. JPOWER - A software package for solving electrical power flow and optimal power flow problems. URL https:// github.com/rwl/JPOWER.

[153] R. D. Zimmerman, C. E. Murillo-Sánchez, and R. J. Thomas. MATPOWER: Steady-state operations, planning, and analysis tools for power systems research and education. *IEEE Transactions on Power Systems*, 26(1):12–19, 2011. doi:10.1109/TPWRS.2010.2051168.

[154] D. H. Wolpert and W. G. Macready. No free lunch theorems for optimization. *IEEE Transactions on Evolutionary Computation*, 1(1):67–82, 1997. ISSN 1089-778X. doi:10.1109/4235.585893.

[155] Sydney Coordinated Adaptive Traffic System (SCATS). An introduction to the new generation SCATS 6, 2012. URL http://www.scats.com.au/files/an_introduction_to_ scats_6.pdf.

[156] D. P. Bertsekas. *Dynamic programming and optimal control.* Athena Scientific, Belmont, Mass., 4th edition, 2012. ISBN 1886529086.

[157] R. Bellman and S. Dreyfus. *Applied dynamic programming.* Princeton University Press, Princeton, New Jersey, 1962. ISBN 978-0-691-07913-4.

[158] A. Fotouhi, D. Auger, K. Propp, S. Longo, and M. Wild. A review on electric vehicle battery modelling: From Lithium-ion toward Lithium-Sulphur. *Renewable and Sustainable Energy Reviews*, 56:1008–1021, 2016. doi:10.1016/j.rser.2015.12.009.

[159] S. Mousavi and M. Nikdel. Various battery models for various simulation studies and applications. *Renewable and Sustainable Energy Reviews*, 32:477–485, 2014. doi:10.1016/j.rser.2014.01.048.

[160] X. Hu, S. Li, and H. Peng. A comparative study of equivalent circuit models for Li-ion batteries. *Journal of Power Sources*, 198:359–367, 2012. doi:10.1016/j.jpowsour.2011.10.013.

[161] A. Seaman, T.-S. Dao, and J. McPhee. A survey of mathematics-based equivalent-circuit and electrochemical battery models for hybrid and electric vehicle simulation. *Journal of Power Sources*, 256:410–423, 2014. doi:10.1016/j.jpowsour.2014.01.057.

[162] M. Ecker, N. Nieto, S. Käbitz, J. Schmalstieg, H. Blanke, A. Warnecke, and D. Sauer. Calendar and cycle life study of Li(NiMnCo)O$_2$-based 18650 lithium-ion batteries. *Journal of Power Sources*, 248:839–851, 2014. doi:10.1016/j.jpowsour.2013.09.143.

[163] C. Weng, J. Sun, and H. Peng. A unified open-circuit-voltage model of lithium-ion batteries for state-of-charge estimation and state-of-health monitoring. *Journal of Power Sources*, 258:228 – 237, 2014. ISSN 0378-7753. doi:10.1016/j.jpowsour.2014.02.026.

[164] D. Pelzer, D. Ciechanowicz, and A. Knoll. Energy arbitrage through smart scheduling of battery energy storage considering battery degradation and electricity price forecasts. In *IEEE PES Innovative Smart Grid Technologies, Asia (ISGT)*, pages 472–477, 2016. doi:10.1109/ISGT-Asia.2016.7796431.

[165] L. Fu, K. Endo, K. Sekine, T. Takamura, Y. Wu, and H. Wu. Studies on capacity fading mechanism of graphite anode for Li-ion battery. *Journal of Power Sources*, 162(1):663–666, 2006. ISSN 03787753. doi:10.1016/j.jpowsour.2006.02.108.

[166] J. Vetter, P. Novák, M. Wagner, C. Veit, K.-C. Möller, J. Besenhard, M. Winter, M. Wohlfahrt-Mehrens, C. Vogler, and A. Hammouche. Ageing mechanisms in lithium-ion batteries. *Journal of Power Sources*, 147(1-2):269–281, 2005. ISSN 03787753. doi:10.1016/j.jpowsour.2005.01.006.

[167] J. Li, E. Murphy, J. Winnick, and P. Kohl. Studies on the cycle life of commercial lithium ion batteries during rapid charge-discharge cycling. *Journal of Power Sources*, 102(1-2):294–301, 2001. doi:10.1016/S0378-7753(01)00821-7.

[168] M. Broussely, S. Herreyre, P. Biensan, and P. Kasztejna. Aging mechanism in Li ion cells and calendar life predictions. *Journal of Power Sources*, 98:0–8, 2001. doi:10.1016/S0378-7753(01)00722-4.

[169] I. Fernández, C. Calvillo, A. Sánchez-Miralles, and J. Boal. Capacity fade and aging models for electric batteries and optimal charging strategy for electric vehicles. *Energy*, 60:35–43, 2013. doi:10.1016/j.energy.2013.07.068.

[170] D. Abraham, J. Knuth, D. Dees, I. Bloom, and J. Christophersen. Performance degradation of high-power lithium-ion cells - electrochemistry of harvested electrodes. *Journal of Power Sources*, 170(2):465–475, 2007. ISSN 03787753. doi:10.1016/j.jpowsour.2007.03.071.

[171] G. Zhang, C. E. Shaffer, C.-Y. Wang, and C. D. Rahn. In-situ measurement of current distribution in a Li-ion cell. *Journal of the Electrochemical Society*, 160(4):A610–A615, 2013. ISSN 0013-4651. doi:10.1149/2.046304jes.

[172] D. Abraham, E. Reynolds, E. Sammann, A. Jansen, and D. Dees. Aging characteristics of high-power lithium-ion cells with $(LiNi_{0.8}Co_{0.15}Al_{0.05}O_2)$ and $(Li_{4/3}Ti_{5/3}O_4)$ electrodes. *Electrochimica Acta*, 51(3):502–510, 2005. ISSN 00134686. doi:10.1016/j.electacta.2005.05.008.

[173] P. Ramadass, B. Haran, R. White, and B. N. Popov. Capacity fade of Sony 18650 cells cycled at elevated temperatures Part I: Cycling performance. *Journal of Power Sources*, 112:606–613, 2002. doi:10.1016/S0378-7753(02)00473-1.

[174] W. Bögel, J. Büchel, and H. Katz. Real-life EV battery cycling on the test bench. *Journal of Power Sources*, 72(1):37–42, 1998. doi:10.1016/S0378-7753(97)02775-4.

[175] J. Belt, V. Utgikar, and I. Bloom. Calendar and PHEV cycle life aging of high-energy, lithium-ion cells containing blended spinel and layered-oxide cathodes. *Journal of Power Sources*, 196(23): 10213–10221, 2011. doi:10.1016/j.jpowsour.2011.08.067.

[176] J. Jaguemont, L. Boulon, and Y. Dubé. A comprehensive review of lithium-ion batteries used in hybrid and electric vehicles at cold temperatures. *Applied Energy*, 164:99–114, 2016. doi:10.1016/j.apenergy.2015.11.034.

[177] G. Ning, B. Haran, and B. Popov. Capacity fade study of lithium-ion batteries cycled at high discharge rates. *Journal of Power Sources*, 117(1-2):160–169, 2003. doi:10.1016/S0378-7753(03)00029-6.

[178] B. Stiaszny, J. Ziegler, E. Krauß, M. Zhang, J. Schmidt, and E. Ivers-Tiffée. Electrochemical characterization and post-mortem analysis of aged $LiMn_2O_4$-NMC/graphite lithium ion batteries Part II: Calendar aging. *Journal of Power Sources*, 258:61–75, 2014. doi:10.1016/j.jpowsour.2014.02.019.

[179] I. Bloom, B. Cole, J. Sohn, S. Jones, E. Polzin, V. Battaglia, G. Henriksen, C. Motloch, R. Richardson, T. Unkelhaeuser, D. Ingersoll, and H. Case. An accelerated calendar and cycle life study of Li-ion cells. *Journal of Power Sources*, 101(2): 238–247, 2001. doi:10.1016/S0378-7753(01)00783-2.

[180] J. Belt, C. Ho, C. Motloch, T. Miller, and T. Duong. A capacity and power fade study of Li-ion cells during life cycle testing. *Journal of Power Sources*, 123(2):241–246, 2003. doi:10.1016/S0378-7753(03)00537-8.

[181] J. Shim and K. Striebel. Characterization of high-power lithium-ion cells during constant current cycling: Part I. cy-

cle performance and electrochemical diagnostics. *Journal of Power Sources*, 122(2):188–194, 2003. doi:10.1016/S0378-7753(03)00351-3.

[182] M. Dubarry, C. Truchot, B. Y. Liaw, K. Gering, S. Sazhin, D. Jamison, and C. Michelbacher. Evaluation of commercial lithium-ion cells based on composite positive electrode for plug-in hybrid electric vehicle applications. *Journal of Power Sources*, 196(23):10336–10343, 2011. ISSN 03787753. doi:10.1016/j.jpowsour.2011.08.078.

[183] S. Käbitz, J. Gerschler, M. Ecker, Y. Yurdagel, B. Emmermacher, D. André, T. Mitsch, and D. Sauer. Cycle and calendar life study of a graphite|LiNi$_{1/3}$Mn$_{1/3}$Co$_{1/3}$O$_2$ Li-ion high energy system. Part A: Full cell characterization. *Journal of Power Sources*, 239:572–583, 2013. doi:10.1016/j.jpowsour.2013.03.045.

[184] I. Bloom, S. Jones, E. Polzin, V. Battaglia, G. Henriksen, C. Motloch, R. Wright, R. Jungst, H. Case, and D. Doughty. Mechanisms of impedance rise in high-power, lithium-ion cells. *Journal of Power Sources*, 111(1):152–159, 2002. doi:10.1016/S0378-7753(02)00302-6.

[185] H. Ploehn, P. Ramadass, and R. White. Solvent diffusion model for aging of lithium-ion battery cells. *Journal of the Electrochemical Society*, 151(3):A456–A462, 2004. doi:10.1149/1.1644601.

[186] R. Wright, C. Motloch, J. Belt, J. Christophersen, C. Ho, R. Richardson, I. Bloom, S. Jones, V. Battaglia, G. Henriksen, T. Unkelhaeuser, D. Ingersoll, H. Case, S. Rogers, and R. Sutula. Calendar- and cycle-life studies of advanced technology development program generation 1 lithium-ion batteries. *Journal of Power Sources*, 110(2):445–470, 2002. doi:10.1016/S0378-7753(02)00210-0.

[187] B. Liaw, E. Roth, R. Jungst, G. Nagasubramanian, H. Case, and D. Doughty. Correlation of Arrhenius behaviors in power and capacity fades with cell impedance and heat generation in cylindrical lithium-ion cells. *Journal of Power Sources*, 119-121: 874–886, 2003. doi:10.1016/S0378-7753(03)00196-4.

[188] T. Yoshida, M. Takahashi, S. Morikawa, C. Ihara, H. Katsukawa, T. Shiratsuchi, and J.-I. Yamaki. Degradation mechanism and life prediction of lithium-ion batteries. *Journal of the Electrochemical Society*, 153(3):A576–A582, 2006. doi:10.1149/1.2162467.

[189] E. Thomas, I. Bloom, J. Christophersen, and V. Battaglia. Statistical methodology for predicting the life of lithium-ion cells via accelerated degradation testing. *Journal of Power Sources*, 184(1):312–317, 2008. doi:10.1016/j.jpowsour.2008.06.017.

[190] J. Schmalstieg, S. Käbitz, M. Ecker, and D. U. Sauer. A holistic aging model for Li(NiMnCo)O$_2$ based 18650 lithium-ion batteries. *Journal of Power Sources*, 257(0):325 – 334, 2014. ISSN 0378-7753. doi:10.1016/j.jpowsour.2014.02.012.

[191] A. Millner. Modeling lithium ion battery degradation in electric vehicles. In *IEEE Conference on Innovative Technologies for an Efficient and Reliable Electricity Supply*, pages 349–356, 2010. doi:10.1109/CITRES.2010.5619782.

[192] O. Erdinc, B. Vural, and M. Uzunoglu. A dynamic lithium-ion battery model considering the effects of temperature and capacity fading. In *International Conference on Clean Electrical Power*, pages 383–386, 2009. doi:10.1109/ICCEP.2009.5212025.

[193] A. Hoke, A. Brissette, D. Maksimović, A. Pratt, and K. Smith. Electric vehicle charge optimization including effects of lithium-ion battery degradation. In *IEEE Vehicle Power and Propulsion Conference*, 2011. doi:10.1109/VPPC.2011.6043046.

[194] R. P. Brent. An algorithm with guaranteed convergence for finding a zero of a function. In *Algorithms for Minimization without Derivatives*. Dover Publications, Mineola, New York, 1973. ISBN ISBN 0-13-022335-2.

[195] J. Neubauer and A. Pesaran. The ability of battery second use strategies to impact plug-in electric vehicle prices and serve utility energy storage applications. *Journal of Power Sources*, 196(23):10351–10358, 2011. ISSN 03787753. doi:10.1016/j.jpowsour.2011.06.053.

[196] S. Bashash, S. Moura, J. Forman, and H. Fathy. Plug-in hybrid electric vehicle charge pattern optimization for energy cost and battery longevity. *Journal of Power Sources*, 196(1):541–549, 2011. doi:10.1016/j.jpowsour.2010.07.001.

[197] S. Saxena, C. Le Floch, J. Macdonald, and S. Moura. Quantifying EV battery end-of-life through analysis of travel needs with vehicle powertrain models. *Journal of Power Sources*, 282: 265–276, 2015. doi:10.1016/j.jpowsour.2015.01.072.

[198] H. Qian, J. S. Lai, J. Zhang, and W. Yu. High-efficiency bidirectional AC-DC converter for energy storage systems. In *IEEE Energy Conversion Congress and Exposition*, pages 3224–3229, 2010. doi:10.1109/ECCE.2010.5618283.

[199] E. M. Krieger. *Effects of variability and rate on battery charge storage and lifespan*. PhD thesis, Princeton University, 2013.

[200] J. Madouh, N. Ahmed, and A. Al-Kandari. Advanced power conditioner using sinewave modulated buck-boost converter cascaded polarity changing inverter. *International Journal of Electrical Power and Energy Systems*, 43(1):280–289, 2012. doi:10.1016/j.ijepes.2012.05.002.

[201] A. Purvins and M. Sumner. Optimal management of stationary lithium-ion battery system in electricity distribution grids.

Journal of Power Sources, 242:742 – 755, 2013. ISSN 0378-7753. doi:10.1016/j.jpowsour.2013.05.097.

[202] Bundesverband der Energie- und Wasserwirtschaft. URL http://www.bdew.de.

[203] C. Casals, A. García, and G. Benítez. *A cost analysis of electric vehicle batteries second life businesses*. Lecture Notes in Management and Industrial Engineering. Springer, Cham, 2016. ISBN 978-3-319-26457-8. doi:10.1007/978-3-319-26459-2_10.

[204] Bundesministerium für Verkehr und digitale Infrastruktur (BMVI). Verkehr in zahlen 2016/2017, 2017.

[205] Land Transport Authority (LTA). Singapore land transport statistics in brief, 2015. URL https://www.lta.gov.sg/content/dam/ltaweb/corp/PublicationsResearch/files/FactsandFigures/Statistics%20in%20Brief%202015%20FINAL.pdf.

[206] U. S. Environmental Protection Agency and U.S. Department of Energy. URL http://www.fueleconomy.gov.

[207] Independent Electricity System Operator (IESO). URL http://www.ieso.ca.

[208] New York Independent System Operator (NYISO). URL http://www.nyiso.com.

[209] Energy Market Company (EMC). URL https://www.emcsg.com.

[210] Pennsylvania-New Jersey-Maryland Interconnection (PJM). URL http://www.pjm.com.

[211] W. Braff, J. Mueller, and J. Trancik. Value of storage technologies for wind and solar energy. *Nature Climate Change*, 6(10): 964–969, 2016. doi:10.1038/nclimate3045.

[212] F. Graves, T. Jenkin, and D. Murphy. Opportunities for electricity storage in deregulating markets. *The Electricity Jour-*

nal, 12(8):46 – 56, 1999. ISSN 1040-6190. doi:10.1016/S1040-6190(99)00071-8.

[213] EPRI. EPRI-DOE handbook of energy storage for transmission and distribution applications. Technical report, Electric Power Research Institute, US Department of Energy, 2003.

[214] J. Eyer, J. Iannucci, and G. Corey. Energy storage benefits and market analysis handbook: A study for the doe energy storage systems program. Technical report, Sandia National Laboratories, 2004.

[215] F. Figueiredo, P. C. Flynn, and E. T. Cabral. The economics of energy storage in 14 deregulated power markets. *Energy Studies Review*, 14:131–152, 2006. doi:10.15173/esr.v14i2.494.

Publications

Journals

- D. Ciechanowicz, D. Pelzer, B. Bartenschlager, and A. Knoll. A modular power system planning and power flow simulation framework for generating and evaluating power network models. *IEEE Transactions on Power Systems*, 32(3):2214–2224, May 2017. ISSN 0885-8950. doi:10.1109/TPWRS.2016.2602479.

- V. Viswanathan, D. Zehe, J. Ivanchev, D. Pelzer, A. Knoll, and H. Aydt. Simulation-assisted exploration of charging infrastructure requirements for electric vehicles in urban environments. Journal of Computational Science, 12:1–10, 2016. ISSN 1877-7503. doi:10. 1016/j.jocs.2015.10.012.

- D. Pelzer, J. Xiao, D. Zehe, M.H. Lees, A.C. Knoll, and H. Aydt. A partition-based match making algorithm for dynamic ridesharing. *IEEE Transactions on Intelligent Transportation Systems*, 16(5):2587–2598, Oct 2015. ISSN 1524-9050. doi:10.1109/TITS.2 015.2413453.

- D. Pelzer, D. Ciechanowicz, H. Aydt, and A. Knoll. A price-responsive dispatching strategy for vehicle-to-grid: An economic evaluation applied to the case of Singapore. *Journal of Power Sources*, 256(0):345–353, 2014. ISSN 0378-7753. doi:10.1016/j.jpo wsour.2014.01.076.

Book Chapters

- D. Ciechanowicz, A. Knoll, Patrick Osswald, and D. Pelzer. Towards a business case for vehicle-to-grid - Maximizing Profits in ancillary service markets. In *Plug In Electric Vehicles in Smart Grids*. Springer, Singapore, 2015. ISBN 978-981-287-301-9. doi:1 0.1007/978-981-287-302-6_8.

© Springer Fachmedien Wiesbaden GmbH, part of Springer Nature 2019
D. Pelzer, *A Modular Framework for Optimizing Grid Integration of Mobile and Stationary Energy Storage in Smart Grids*, https://doi.org/10.1007/978-3-658-27024-7

Conferences

- R. Bi, J. Xiao, D. Pelzer, D. Ciechanowicz, D. Eckhoff, and A. Knoll. A simulation-based heuristic for city-scale electric vehicle charging station placement. In *IEEE International Conference on Intelligent Transportation Systems (ITSC)*, 2017. doi:10.1109/ITSC.2017.8317680.

- D. Pelzer, D. Ciechanowicz, and A. Knoll. Energy arbitrage through smart scheduling of battery energy storage considering battery degradation and electricity price forecasts. In *IEEE Innovative Smart Grid Technologies Conference, Asia (ISGT)*, pages 472–477, Nov 2016. doi:10.1109/ISGT-Asia.2016.7796431.

- D. Ciechanowicz, D. Pelzer, and A. Knoll. Simulation-based approach for investigating the impact of electric vehicles on power grids. In *IEEE PES Asia-Pacific Power and Energy Engineering Conference (APPEEC)*, Nov 2015. doi:10.1109/APPEEC.2015.7381046.

- M. Wagner, V. Viswanathan, D. Pelzer, M. Berger, and H. Aydt. Cellular automata-based anthropogenic heat simulation. In *International Conference On Computational Science (ICCS)*, volume 51, pages 2107–2116, 2015. doi:10.1016/j.procs.2015.05.480.

- D. Pelzer, M. Peters, H. Hauser, M. Rüdiger, and B. Bläsi. Diffractive structures for advanced light trapping in silicon solar cells. In *32rd Workshop of the European Society for Quantum Solar Energy Conversion*, 2011.

- M. Peters, M. Rüdiger, D. Pelzer, H. Hauser, M. Hermle, and B. Bläsi. Electro-optical modelling of solar cells with photonic structures. In *25th European PV Solar Energy Conference and Exhibition (PVSEC)*, pages 87-91, 2010. doi:10.4229/25thEUPVSEC2010-1AO.6.6.

Printed in the United States
By Bookmasters